Praise for

The Boy Who Felt Too Much

"This is a beautiful book. It will help the neurodiverse movement give these beautiful kids a bright and positive future and will have a global impact on how they are viewed, educated, and treated. It will shape their impact on the world."

—Henry Markram

"Wagner does an exceptional job of mixing biography with complex yet accessible neuroscience. Essential for parents, educators, physicians, and specialists working with children of all ages."

—*Library Journal*, starred review

"Lorenz Wagner's book on the remarkable Markrams is, well, remarkable. This is an extraordinary story of noted neuroscientists seeing the world through the eyes of their ASD son and then helping change the way autism is seen and treated far and wide. A tale of love, constancy, and groundbreaking research."

—Ron Suskind, Pulitzer Prize–winning journalist and author of *Life, Animated: A Story of Sidekicks, Heroes, and Autism*

"A compelling story about one of the world's most brilliant researchers searching to understand his son's autism. A gloriously positive book. Left me with hope for humanity."

—David Sinclair, *New York Times* bestselling author; professor in the Department of Genetics and codirector of the Paul F. Glenn Center for the Biology of Aging, Harvard Medical School

"*The Boy Who Felt Too Much* tells the story of our accelerating understanding of autism, and it does so via fascinating characters, such as the famous neuroscientist Henry Markram, who is building a digital brain, and Dr. Markram's remarkable autistic son, Kai. It offers a very enjoyable way for anyone to become fluent in the latest theories about autism, including why it seems so prevalent today and what can be done to minimize its most troublesome aspects. This book greatly increased my respect for the condition of autism, as well as the individuals and families who navigate life with it."

—Martine Rothblatt, chief executive officer, United Therapeutics Corp

"What is so special about this book? It brings together two worlds. Many parents will find themselves in the wonderfully, movingly told family story. At the same time, the book delves deeply into brain research. It is very competently researched, based on studies and clearly and visually told, so that everyone can understand the facts. The book is now on everyone's lips among our parents. It offers encouragement in many places and should be read from our point of view by many parents."

—*AUTISM* (journal of the Federal Autism Association of Germany)

"A moving father-son story interspersed combined with elegant science writing about regions of the human brain that are still terra incognita. Thoroughly researched and brilliantly written."

—*Weser Kurier*

"A book for all those dealing with autism and those who love good writing."

—*Süddeutsche Zeitung*

"By the end of this striking piece of reportage, you've followed not only an unusual family story but also the rigorous, deliberate, but extremely vivid search for knowledge."

—*Der Falter*

"A wonderful book."

—*Neue Zürcher Zeitung am Sonntag*

"A thrilling and moving book."

—*WDR 5*

"Thank you for this really interesting and beautiful book."

—*SR2*

"A family portrait with amazing insights."

—*Technology Review*

"Lorenz Wagner has written an impressive portrait of the family with a focus on the father and has succeeded in making highly complex processes understandable to laypeople. A very interesting read."

—*Reutlinger General-Anzeiger*

"Woven in this lively story, Wagner explains what happens in us when we feel and provides fascinating insights into the human brain. A captivating and uplifting book."

—*Bild der Wissenschaft*

"The author is really good writer. Wagner interweaves Markram's biography with intimate family sketches and passages on neuroscience and autism research."

—*Saarbrücker Zeitung*

"The first time autism has been shown from another side."

—*Asperger-Autismus.ch*

"Thank you for this great and extremely important book."

—Ella's Blog, *Living with Autism*

"Wonderfully visual. There are few books that can provide both personal and scientific insight into the autism spectrum. This book is able to fill the arc of tension skillfully, excitingly, and at the same time in an entertaining fashion."

—*Atypical life* autism blog

"The book is very moving and exciting and can be read like a novel. A brain researcher discovers that his child is autistic, and as readers we see how he gradually learns to deal with it. A journey into a new world."

—*Familylab*

the boy who *felt too much*

the boy who *felt too much*

HOW A RENOWNED NEUROSCIENTIST AND HIS SON CHANGED OUR VIEW OF AUTISM FOREVER

LORENZ WAGNER

Translated from the German by Leon Dische Becker

ARCADE PUBLISHING • NEW YORK

First English-language Paperback Edition 2021

Arcade Publishing books may be purchased in bulk at special discounts for sales promotion, corporate gifts, fund-raising, or educational purposes. Special editions can also be created to specifications. For details, contact the Special Sales Department, Arcade Publishing, 307 West 36th Street, 11th Floor, New York, NY 10018 or arcade@skyhorsepublishing.com.

Arcade Publishing® is a registered trademark of Skyhorse Publishing, Inc.®, a Delaware corporation.

Visit our website at www.arcadepub.com.
Visit the author's website at lorenzwagner.com.

10 9 8 7 6 5 4 3 2 1

Library of Congress Cataloging-in-Publication Data

Names: Wagner, Lorenz, author.
Title: The boy who felt too much : how a renowned neuroscientist and his son changed our view of autism forever / Lorenz Wagner ; translated from the German by Leon Dische Becker.
Other titles: Junge, der zu viel fühlte. English
Description: First English-language edition. | New York : Arcade Publishing, [2019]
Identifiers: LCCN 2019026513 (hardcover) | LCCN 2019026514 (ebook) | ISBN 9781948924788 (hardcover) | ISBN 978-1-951627-49-2 (paperback) | ISBN 9781948924795 (ebook)
Subjects: LCSH: Markram, Henry. | Parents of autistic children—Germany—Biography. | Autistic children—Germany—Family relationships—Biography. | Fathers and sons. | Brain—Research.
Classification: LCC RJ506.A9 W33313 2019 (print) | LCC RJ506.A9 (ebook) | DDC 618.92/858820092 [B]—dc23
LC record available at https://lccn.loc.gov/2019026513

Cover design by Erin Seaward-Hiatt
Cover photograph courtesy of the Markram family

Printed in the United States of America

For Romy

Contents

I
THE MYSTERY

1
Is That Your Kid?

"Is that your son?"
"Yeah, why?"
"You won't believe what he just did."

The car was coasting. Kai heard the wheels crunch as it drew to a halt outside his house. The car door opened, and a young man hopped out. He popped the hood and disappeared beneath it. "You've got to be kidding me!" he fumed.

Kai emerged from the front yard. It was late in the morning, and the street was empty. Cars were a rare sight around there. Kai often played in the street with his sisters. Mostly only bikes passed by, anyway—students on their way to class. Kai lived with his parents and sisters on a sprawling college campus, which boasted sculptures, water fountains, benches, flame trees, a Japanese garden, and enough space to spend the whole day wandering idly, accompanied by chirping birds.

"Hello. I'm Kai."

The man ignored him.

"Is your car not working?"

"No," the man grumbled. How would he get to class now? He was stuck in this damn residential area and would be too late for his exam. If he didn't make it there on time, they'd mark him absent and fail him.

Kai spun around and dashed off. The man got back in his car and turned the key. The engine sputtered for a moment and died again.

The boy came running back. What the hell did he want?

He was holding something in his hand.

"Here," said Kai. "My mom's key."

"Excuse me?"

"You can take our car."

The man stared in disbelief and took the key.

* * *

Kai loved people, and it was hard not to love him back. At the age of two, he started twisting his way out of his father's grip and running over to them—random pedestrians, the postman, elderly people sitting on benches basking in the warmth of the morning sun. Kai opened his arms and clung to their legs. He didn't say anything. They froze at first, but when they looked down to see his twinkling brown eyes staring up at them, they couldn't help but laugh. Kai didn't speak much. He spoke with his hands and glowed from inside. He warmed their hearts more than the sun possibly could. Soon they were sitting on the benches because of him, the little boy who had just moved to Rehovot, Israel.

Kai was born in Heidelberg, Germany, on June 21, 1994, the first day of summer. It was the longest day of the year and proved to be the longest birth his mother Anat would have to endure, dragging on some twenty hours. While she twisted and turned in pain, Henry wandered up and down the hallway. Their two daughters, Kali and Linoy, would soon have a baby brother. They could barely contain their excitement.

The midwife laughed when she held Kai up by his feet. He was so long and heavy and had so much hair on his head. "We might as well put a jacket and pants on him," she said. "Send him straight to kindergarten."

* * *

Babies are often born with a smile on their face. Experts call this a *reflex smile*. It's a newborn's way of ingratiating itself. That smile is many parents' first memory of their child. Henry doesn't remember if Kai was smiling. He remembers something else: newborn Kai kept trying to lift his little head. His wide eyes had this absorbing glimmer, tracking every light and sound, darting back and forth on high alert.

Henry had treated many babies during his time at medical school, but he had never seen eyes like that. His son's gaze seemed targeted, intentional. That was impossible. A newborn's vision doesn't develop for another few months. Until then, everything is blurry to their eyes—colors, contours. They can only see what's right in front of them: their parents' faces, the mother's breast. Kai, however, behaved as though he could see.

His pupils darted nonstop. Henry was worried. The doctors on the ward huddled. They had never seen a child like that either. As they inspected Kai carefully, the worries vanished from their faces. Kai scored a full ten points on the Apgar test, which rates a newborn by appearance, pulse, grimace, activity, and respiration. "All good, boss," his colleagues told him. Henry's fears turned to pride: "He is the most alert child on the ward," he told Anat. "Our son is something special."

Anat wasn't all that comforted. She watched Kai even more closely. When he was around six months old, she noticed a change in his eyes. She couldn't put it into words; it was a feeling more than anything. Henry didn't see it. Neither did the doctor they consulted. "A wonderful child," he assured them, "fit as a fiddle."

"You see?" Henry said to Anat. "Everything's fine."

And so, the Markrams' life continued blissfully. Kai's bassinet became a stroller, then a tricycle. Their house echoed with laughter and happy shrieking. They spoke and joked in multiple languages—English, Hebrew, German. Henry was from South Africa, Anat from Israel. His work as a neuroscientist had brought them to Heidelberg. He had made a name for himself in his field at an early age, asking questions and devising answers that seemed precocious for a grad student. Bert Sakmann, a German Nobel Prize winner in medicine, recruited him to the Max Planck Institute for Medical Research. Who knows? Maybe this Markram would someday win the Nobel Prize himself.

His family loved Heidelberg, the colorful houses, the winding alleys, the Neckar River, the famous castle. On weekends, they drove out to the countryside, went swimming and ice-skating, picked apples and asparagus. They spent their vacations travelling in Europe, which was entirely new territory for them. Paris, Rome, Copenhagen: Henry carried Kai around in a blanket, with the girls skipping alongside them and Anat taking photographs. It was the best time. All their worries awaited beyond the horizon.

* * *

They had spent two years in Heidelberg when Henry got the fateful call from Israel. By then, the postdoc had lived up to the promise his mentor Sakmann had seen in him. Henry had researched how brain cells communicate with each other, even inventing a way to watch them do it, a method that would soon be standard in labs around the world. At the young age of thirty-five, he landed a job at the prestigious Weizmann Institute in Rehovot, where he would become a professor, build a laboratory, and lead his own research department.

Kai had grown up to be a happy child. Wild curls swirled around his head; his eyes were much too big for his face. When he laughed, his nose wrinkled. When he spoke, you could see the gap between his front teeth. He would often say things that were beyond his years. He was a tad professorial for a toddler. "What a special boy," the neighbors said.

Henry and Anat watched him in awe, moved and amused by their little boy. Kai was an enigma. He didn't speak much. He said only what was necessary. And greeting strangers was absolutely necessary. He greeted every person he saw. "Hello. I'm Kai." If they responded in kind or just smiled, Kai would remember their face and what they were wearing and add them to his growing list of friends. Referring to these encounters days later, shocked that his parents couldn't recall them, Kai would jog their memories with minor details: he meant the woman with the pink flowers on her hat, the man with the smudge of dirt on his boot. He raised his voice, his cheeks glowing red: How could they not remember?

If they ever said something snarky about one of his friends, say, that the flowers on the woman's hat were too bright and pink for their taste, Kai would burst into tears and shriek, "You shouldn't say that!" Henry and Anat smiled. He was right, of course.

His kindergarten counselors would never forget him. Years later, their hair now gray, they still talk about young Kai, how he would mosey from table to table, arms folded behind his back like an old gentleman. He didn't draw; he preferred to watch others drawing. When he wanted to play with other children, he didn't ask, he just reached out and touched them. Sometimes his gesture came as a surprise or his grip was too tight. The other kids thought he was trying to push them and shoved him back. Kai was aghast, but he didn't cry, even if they bruised or scratched him. Kai wasn't good at sharing his suffering.

His sisters were gentle with him. They accepted Kai as he was. But sometimes they did wonder. They loved feeling just a bit creeped out when their parents read them bedtime stories, but their brother lost it completely. A reading of Goldilocks was too much for him. "Stop!" he would scream, running out of the room and slamming the door behind him. To Kai, this wasn't just a story—it was real. After a tearful night caused by the death of Bambi's mother, the family agreed that it would be best for all involved if they only read *good night* stories that merited that designation.

There would be countless such experiences. No child in the neighborhood was discussed, laughed about, or puzzled over more than Kai. "Kids are kids," Henry would say. Every one of them lives in their own world. Wasn't it great that Kai had an imagination, an eye for detail, that he loved everyone and approached them?

Sitting by the open window, Anat could often hear Kai in the garden accosting pedestrians. "Want to come in and have coffee with my mom?" Luckily for her, sadly for Kai, they usually turned him down, and Anat remained undisturbed in their messy kitchen.

But today was different. The doorbell rang and Anat answered it. A young man stood in the doorway in front of her. No, she didn't want to buy anything.

"Is this your key?"

"Excuse me?"

"Your son gave it to me."

"What?"

"My car broke down, and your son—"

"Kai!"

Five minutes later, the three of them were sitting in the car. Anat drove, the student kept an eye on his watch. He would make it after all.

"What would I have done without your help?" he said.

"You have Kai to thank for that."

"What a special boy."

Anat nodded.

2
The Boy Who Changed Everything

If Henry had just been a scientist,
even a great one,
he would have failed.

Kai is different. Kai, the doctors would eventually realize, is autistic. Of course, like everyone on the spectrum, Kai isn't just autistic—he's so much more than that. Kai is Kai.

Doctors used to find one case of autism among every five thousand people. Today, according to a study by the US Department of Health & Human Services, the ratio is one in fifty-nine. Scientists speak of an epidemic. Kai may be different, but he is not alone.

Henry is one of the world's most famous neuroscientists, but when Kai began withdrawing, he was as helpless as all the other parents of autistic children. He asked himself the same questions they did: What is autism? How can I help my child?

He researched for fifteen years. His findings would upend everything we thought we knew about autism, and offer us a new way of thinking about a number of conditions.

If Henry had just been a scientist, even a great one, he would have failed. He only succeeded thanks to Kai, the boy who changed everything.

* * *

It's 4:00 a.m. Henry throws off the bedcovers. He tiptoes out of the bedroom, across the hallway into the kitchen and makes coffee. Quietly. Everyone is sleeping. He opens his laptop. The bluish light of the screen shines on his face. His eyes are smaller than usual, his hair messy. He is thin. Only a few weeks ago, he was in Portugal on a fasting retreat. He slurps his coffee and reads e-mails.

"Dear Henry," a lady called Sandra writes to him. "I am autistic. Reading your story, I was overcome by emotion. For the first time in my life, someone was describing my experience. My family doesn't support me."

"Dear Sandra," Henry types. "I know what you're going through."

He reads more e-mails from autistic people, their families, his colleagues. He reviews the data—rows of numbers that only a scientist can understand. Finally, he opens a lecture he was working on until midnight. "We think we see with our eyes," it says, channeling Antoine de Saint-Exupéry's *The Little Prince*. Unlike the Little Prince, however, Henry doesn't believe we see with our hearts: our brains shape our view of the world.

"The pathways in your brain are so extensive, you could stretch them once around the whole moon: a hundred billion brain cells, a hundred billion synapses, a wonderful system, and six hundred ways to disturb it. Autism, ADHD, depression, Alzheimer's, Parkinson's, schizophrenia. How are they all related?"

This question drives Henry out of bed every morning at 4:00 a.m. He is certain it will be answered in our lifetime. Humanity will decode the brain. And then rebuild it. Henry plans to do the rebuilding himself. He kicked off that project ten years ago. The European Union earmarked it with a billion-euro grant. It may lead to the greatest scientific achievement in history, greater than the decoding of the genome, greater than the moon landing. Humankind would finally understand itself.

Will he, the guy from the Kalahari Desert, be one of those scientists who make history?

* * *

Henry grew up in South Africa, spending his childhood on his grandfather's farm. His family was well-off. They had settled in the Kalahari generations ago, but life was hard. "Nothing comes easily in the desert," his grandfather would say. "You have to work hard for everything." Henry was put to work as soon as he could walk, doing household chores, milking cows. If he wasn't up before sunrise, his grandfather stormed into his room with a whip and threw him out of the house, whipping him out the door.

His grandfather was a Boer. At ninety-five, he could still be seen galloping across the desert with a straight back. He didn't talk much and was hard on everyone, especially himself. Henry's five uncles, who also lived on the farm, hunted with their bare hands. "C'mon Henry," they would summon him and jump into the Jeep. Once Henry was a bit older, ten or eleven, he became their driver. He sat behind the wheel, barefoot, the sun stinging, the dust rising, his uncles peering out toward the horizon.

"There! You see that kudu?"

Henry put the pedal to the metal. Grass and bushes flew by, the wind burned his eyes, the speedometer hit 30 miles per hour. He skidded to a halt alongside the kudu, and one of his uncles jumped on the creature, grabbing it by the horns. A kudu weighs up to 800 pounds, but with all that thrust and grip, its neck was swiftly broken. As the dust settled, his uncles strung up their catch on the Jeep's roof and slaughtered it right there. They drank whiskey as they worked, offering Henry a swig, which he refused.

Today, that feels like someone else's life, Henry says, though he still rises early.

Henry's mother was British. She didn't feel at home in the Kalahari but still saw the good in their lives there. The desert had given her son a childhood in nature: the vast expanses during the day, the stars so very close at night. But as Henry got older and his voice deepened, his mother sensed that he was outgrowing the place. "This world has nothing to offer him anymore," she said. "He will not become a farmer." She sent him off to a private school near Durban on the other side of South Africa. Leaving home was hard on Henry. He missed his family, the landscape, the farm. In time, however, he discovered a love of learning that was just as strong. His visits home became ever more rare.

* * *

Henry had another uncle, who didn't live on the farm. Uncle John was sensitive. He read books. And Henry loved him. On some days, however, John wasn't himself. He hardly spoke, and the look in his eyes was somber and heavy. One day, when Henry was fifteen, his mother took him aside. "Uncle John is dead," she said. "He took his own life." Henry cried. He didn't understand depression, the monster that had taken hold in his uncle's head. He went to the library in search of answers.

The human brain, he read, weighs three pounds. Its texture is soft, like those slabs of liver you can buy at the butcher shop. It's pink and shiny and only turns gray when you die and the cells start to rot. The brain is covered in folds or wrinkles and lined with fine capillaries. It looks so neat and tidy.

The brain runs on electric current. A computer that could perform its tasks would require billions of dollars in electricity. Electric current flickers through your brain nonstop. When you see a rose, it isn't your eyes formulating the image. Your optic nerves fire electric signals, and your brain compiles the information: *red, petals, a stem, thorns = a rose.* Once the image is complete, you evaluate it. It is beautiful. It smells good. Watch out or it might prick you! Your brain allows you to see, think, and also feel. Henry wanted to know all about this. Why had his uncle been so sad? What makes people feel emotions in the first place?

When Henry asked himself that question at the age of fifteen, he had no way of knowing it would come to define his whole life. Or that this same question would not only determine his uncle's fate but also shape the destiny of his unborn son.

Our feelings develop in two parts of the brain, the amygdala and the cerebral cortex. It's simple, really: You see a snake. Your eyes send a signal to the amygdala and it sounds the alarm. Your heart starts beating faster, your blood gets warmer, and your body prepares to fight or take flight. You don't even know you're afraid yet.

Those feelings arise when the signals reach your cerebral cortex, farther down the chain of command. Your sense of reason chimes in. Your cerebral cortex collects the details and evaluates them: Does the snake have fangs? Is it threatening you? The cerebral cortex summons memories. Have you experienced something like this before? Then it sends its verdict to the amygdala: *it's just a snake charmer. You've seen one on television*. Your pulse slows. You relax.

If the pathways between the cerebral cortex and amygdala are disturbed, your feelings will be disturbed as well. For example, experiments have shown that cats become reclusive if you remove their cerebral cortex. If you stroke such a cat, it will hiss at you, and it will have no idea why. It has no feelings. The hissing is just a reflex. Scientists call this *sham rage*.

In the human realm, we have the famous case of Phineas Gage. A construction foreman, Gage worked on the railroads as they expanded across the United States. He loaded blast holes, covered them with sand, and lit the fuse. On one occasion, however, he forgot the sand. As he tried to fasten the charge with his hammer, it exploded in his face, driving a three-foot iron rod through his skull. Dr. John D. Harlow removed the rod, which was sticking out of the top of his patient's head. Gage remained conscious throughout. A few weeks later, he returned to work. He spoke normally. He could still smell, hear, walk, and remember everything. But his personality had changed. The once popular and even-keeled worker had become choleric. He insulted people; he could hardly restrain himself. The rod had penetrated his cerebral cortex. "It has destroyed the area of the brain that produces empathy," Dr. Harlow suggested. Scientists have since confirmed this.

Feelings come from our cerebral cortex and amygdala. They contain our very character. Even minor flaws can disturb their equilibrium.

Hiding away in his growing fortress of books, Henry started to understand that there were reasons for Uncle John's sadness. Rational reasons. The world hadn't made him sad; the feeling was coming from inside his head. Henry found this comforting. There had to be a way to help such people. If only he had known this while his uncle was still alive. He would have told him there was hope. Henry decided to become a doctor, a neurologist. He wanted to understand the electricity and chemistry that determines whether we are happy or sad, and even, sometimes, whether we live or die.

Before long, he was at the top of his class in the natural sciences, and, after graduating from high school, he enrolled at the University of Cape Town, majoring in medicine and psychiatry.

3
The Check

His grandfather called him into his room.
He handed him a check.
"Go," he said.

More than a hundred years ago, a gifted doctor named Paul Ehrlich lived in Berlin. His patients adored him. He always had a good word for them. Where other doctors shrugged, he always had a suggestion that lent them hope. Ehrlich was ahead of his time. He complained publicly that women were paid less than men. And in medicine, he dedicated himself to new and unprecedented methods.

At the time, accelerating industrialization was transforming the Western world. Railroads and assembly lines set the pace. New technologies and new sciences were changing the way people lived: steam power, electricity, physics, and chemistry. Only the medical profession seemed stuck in its ways. The epidemics of the age—tuberculosis, diphtheria, cholera, and syphilis—were still widely viewed as divine judgments. Great bravery was required to shake up the status quo, and doctors like Robert Koch, Emil von Behring, and Ehrlich rose to the occasion. They wanted to introduce the burgeoning sciences into their field. They withdrew from hospital bedsides and went to their laboratories. Drawing on advances in microscopy and chemistry, they discovered that mass epidemics weren't the work of a vengeful God—or, as more secular thinkers would have it, the result of vapors pouring out of holes in the ground—but instead were caused by tiny living things called bacteria.

When Ehrlich withdrew to his lab, his colleagues rolled their eyes at him. They distrusted all this new mumbo jumbo. Chemistry! Staining cells! But he remained hunched over his microscope. Through constant experimentation, he discovered chemicals that made certain tissue cells visible for the very first time. He discovered an unknown form of blood cell, leukocytes. He became an immunologist and, in 1908, he received the Nobel Prize in Medicine.

Ehrlich's hospital colleagues accused him of neglecting the sick. Diphtheria was killing fifty thousand children a year at the time. But Ehrlich no longer saw it as his primary mission to cure diseases. He wanted to prevent them instead. The pathogens shouldn't be allowed to get to work in the first place; the toxins would be rendered harmless by "antitoxins." Ehrlich wanted to vaccinate people.

Another independent spirit, Emil von Behring, built on Ehrlich's work and developed a vaccine, diphtheria antitoxin. It was distributed in vials inscribed "per Behring and Ehrlich." Within a year, child fatalities were reduced by 75 percent.

Ehrlich started experimenting with so-called magic bullets, substances from the realm of chemistry that could kill microbes without harming humans. He tested these pellets on animals. Chemicals in medicine? The elders shook their heads. Ehrlich was banned from the Charité hospital in Berlin. But several years later, his critics had to acknowledge that the "magic bullets" worked, just as antibiotics and chemotherapy would later. Ehrlich revolutionized medicine and founded a new kind of pharmaceutical research. More than a century later, medical science still follows his two main principles: microscopic analysis and animal testing.

* * *

Bo-o-o-ring. Henry couldn't stand it any longer. Day after day, he drove to the university and sat in the stuffy lecture halls, where professors lectured and students listened.

He had learned a lot about the human body and brain in the beginning, and found his internship in the hospital interesting, too, particularly in the pediatric ward. He loved children and wanted five of his own, but that required a wife and a bit of free time, and his whole life up to that point had been dedicated to studying. He got the basics down quickly and felt ready to delve into the nitty-gritty: the structure of the brain and all the exciting new ways it would be explored in the future. At least, that was his hope. What his professors offered instead, on their blackboards and overhead projectors, was a disappointment. How little humanity knew about the brain! How far we were from solving the mysteries of thinking and feeling! And how gutless the attempts had been to change that! The lectures were only marginally concerned with providing actual understanding and certainly didn't encourage further exploration. They just taught you how to correctly apply the old teachings.

So, this was the life of a neurologist: You inquire about the symptoms, pick up the book, find the appropriate entry, and name the disease. You find that there are, say, five medications to treat it. You prescribe the first one. And if it doesn't work, you prescribe the second, the third, the fourth, the fifth. You rarely know why those particular medicines work. In most cases, they don't even promise a cure, just relief from the symptoms. Alzheimer's, Parkinson's, schizophrenia, and autism—all entirely incurable.

Henry started skipping class. He preferred going down the hallway to the laboratory wing, the wing of the explorers. How he loved their world! The labs buzzed and blinked, smelled like formaldehyde, and were giddy with the spirit of inquiry and adventure. Doctoral students and professors immersed themselves in Ehrlich's dyed cells and explored organs and organisms. Henry simply joined in. He greeted them like colleagues, and when they greeted him back, he nodded knowingly, said something smart, and kept coming again and again. Eventually, he was one of them, wearing his own lab coat over his clothes. "Oh," he said, "the gentlemen are isolating nephrons," small functional units of the kidney that filter water in the body. Henry let the nephrons dry like a sponge, added saltwater, and then watched the units drain all salt from the water.

Let someone else become a doctor, he thought. I'll become a scientist instead, a researcher. Like Ehrlich, he wanted to fight the scourges of his time: Parkinson's and ADHD, depression, and autism. Those were the new epidemics, the number of cases growing exponentially year by year. He would write his dissertation on brain cells, how they communicate and understand each other. This question contained humanity's essential mystery: Who are we?

First, he would have to transfer to a research university. Several professors recommended the Weizmann Institute in Israel, one of the most influential research centers in the world. It had already produced three Nobel Prize winners, as well as crucial medicine to fight cancer and multiple sclerosis. Unlike most other universities, it was future-minded, one of the first places with a so-called mainframe computer. Henry applied for a scholarship, and, what do you know, they ended up giving it to him. Him: twenty-six-year-old Henry Markram from the Kalahari. He flew home proudly.

A grand welcome. The boy was back. What a surprise! If he had let them know ahead of time, Mum would have baked a cake. He had gotten so thin. Wasn't he eating enough? You can't work all the time, you know. Your brain needs food too. They sat down for dinner.

"I'm moving to Israel," Henry announced.

The family stopped chewing and chatting. "You're what?"

"I'm going to the Weizmann Institute."

"What?" "Huh?" "Where?" "Israel? Why would you want to go there?" "Isn't it dangerous?" "No, you're not. You're staying here!" They continued in this manner.

"I want to do research," Henry said. "The Weizmann Institute is famous for its research."

"What research?" "We thought you were becoming a doctor! You said you wanted to help people." "Spending all day in the lab. That's no way to live." "And how do you intend to feed your family? The salaries in science aren't exactly up there, you know."

Henry got an earful from his parents. This was not what they expected when they paid his high tuition. They certainly wouldn't support a whim like this. Like so many sons and daughters before and after him, Henry had returned home feeling mature and confident only to be reduced to adolescence in a matter of hours.

Discouraged and gloomy, he sneaked upstairs to his childhood bedroom. Someone called his name. It was his grandfather, who used to drive him out of bed with a whip and never had much to say to him. Grandpa had dedicated his life to the farm. He never left. But now he called Henry into his room and handed him a check. "Go," he said. That was it, but there was love in his voice.

Henry never understood why his grandfather helped him. He never got to ask. When he returned years later as a promising neuroscientist, bestowed with so much early success that the whole family welcomed him proudly—their son, the great scientist—his grandfather was dead. The old man never found out what had become of his investment. Henry never got the chance to tell him how the Weizmann Institute had changed his view of the world. He would have done so with pleasure. He would have told his grandfather all that he had learned about the cerebral cortex, which contains almost everything that makes us who we are: intellect, memory, emotions. Our brain is like the universe, he would have said. It has as many cells as the sky over the savanna has stars.

How a person thinks, feels, and develops is determined by interactions between their neurons. When a child's brain grows, the child isn't growing more neurons but augmenting the connections, the pathways, between them. If those pathways are disturbed, that changes how a child develops.

Neurons transmit everything you hear, see, and feel: that piece of music you were playing on the old stereo system, the light from the lamp on the front porch, the itchy socks your grandmother knitted for you. When you turn on the light, neurons get excited. When you try on those socks, a few neurons may even pop. When the radio plays a piece of music, the neurons become calm, almost depressed. Depending on whether the stimulus is light, touch, or sound, neurons behave differently, sending different messages.

And while the stars over the Kalahari remain beyond our reach, in the universe of the brain, you can intervene. While the music is playing, you can dribble a drop of acetylcholine, a chemical messenger, on a neuron, and that once-depressive neuron will flip out. The way it perceives the world will have changed utterly.

As a scientist, Henry could observe that neuron at work. He had seen it with his own eyes: reality isn't just out there; it's created. Of course, a chair is a chair, but every person has a different view of the world. It's completely subjective whether you perceive it as colorful or pale, bright or dark, happy or sad. Some people see colors more brightly when they're angry; others perceive them as paler. Scientists have yet to understand why that is. All they know is that the objective world is a fantasy. Our coexistence is a muddle of interpretations. We see the world so differently.

Henry's grandfather would probably have laughed out loud and said, "When your uncles drink whiskey, they also see the world differently." Old age lets us see matters more calmly. But another of Henry's revelations would have made his grandfather perk up. Looking into a microscope is a lot like beholding the night sky over the Kalahari. You see a universe of neurons and synapses. You delve into infinitude, you fly into space, surrounded by wild lights, and suddenly the stars start connecting with each other, spinning threads of light that allow them to communicate. When they have something to say to one another, a light flashes. The greatest fireworks light up all around you; the explosions spread through your brain like a swarm of birds. It's a miracle, really.

And thanks to your help, Grandpa, thanks to your check, he would have said, I now know my way around that universe. I can read the landscape the way the bushmen read the Kalahari. When they stand on a mountain and look out over the savanna, they see things we can't see: the future weather, the imminent dangers, a lion sitting a mile away. They see how everything is connected, the big picture. That's how it is when I look through a microscope. I see something that no one else can. I can't see the big picture quite yet. That's still a long way off. His grandfather would have nodded and said, "You can do it."

In expert circles, too, many people thought Henry could do it. His study of two neurotransmitters, and how they could help people focus, caused the scientific community to take note. Henry was invited to the United States by the National Institutes of Health (NIH), the world's largest provider of research grants. After Henry spent a year in Washington, DC, Bert Sakmann brought him to Heidelberg. Under Sakmann, Henry developed his method for measuring how cells communicate with other cells. He won prizes, traveled the world, gave lectures. He was flying high. He seemed unbeatable. And then along came Kai.

4
Evaluating Kai

The doctor studied her.
"Maybe your son isn't the one with the problem," he said.
"Maybe it's you."

Some diseases take hold suddenly. A flu or a virus, a sore throat or stomachache strikes you down, and you feel miserable. But those are rarely the worst ailments.

Other diseases creep up on you. Terrible ones: cancer and heart disease. They also seem to appear out of the blue, but if you're vigilant and lucky, you can catch them early. They have precursors: tumors that grow, blood vessels that clog. They can be screened for or felt by a doctor's knowing hands.

Afflictions of the mind creep up on you even more insidiously. One cannot screen for them or palpate to find them. They remain invisible long after they've developed. They come disguised as a quirk, a tic, or a minor discontent. No one worries when you wash your hands a bit too much, until it proves to be compulsive. No one is alarmed when you're sad, until it turns out to be depression. Everyone laughs when you lose your keys again, until one day you no longer find your way home. These invisible, incomprehensible illnesses are, in some ways, the worst.

Kai's autism sneaked into their lives so quietly and subtly that even Henry, the neuroscientist, and Anat, his vigilant wife, didn't recognize the severity of what was happening to their son. None of the doctors they consulted set them on the right track. They all saw individual symptoms, never the full disorder. They saw puzzle pieces rather than the big picture.

Kai's stunted speech was the first thing that worried them. He was three years old when they settled in Rehovot, yet he hardly spoke. This was counterintuitive because Kai was more outgoing than anyone else in the family. Unfortunately, he still communicated almost exclusively with his hands. People didn't understand that very well, and Kai's attempts at reaching out grew ever more frustrating. Try as he might, he either got no response or a deeply unsatisfying one. His face contorted. He was inconsolable.

"Maybe it's because you speak English to Kai and your wife speaks Hebrew," one doctor said.

What nonsense, Henry thought.

They started looking for a therapist. They went to one, then another, then a third, but Kai, otherwise happy and friendly, couldn't stand any of them. He looked at the men in their white coats and fell silent or just walked out of the room. Finally, they found a doctor he liked, who was willing to treat him. She practiced words and sentences with Kai, and his tongue started moving, even if he still garbled syllables. *Here we go*, Henry thought.

Language delay is a sign of autism, but no one considered that possibility.

Kai started covering his ears. His family took him to the movies to see *The Little Mermaid*, but Kai blocked out the noise and wanted to leave immediately. The next time, he refused to go altogether. "That doesn't mean anything," said the doctors. "He's just a bit sensitive."

He was clumsy too. Kai stumbled more often than either of his sisters had at his age. He climbed up a bookshelf and tumbled off backwards. He ran through a windowpane, blood spewing everywhere, and had to get stitches. Again and again, he scratched the scar open. Here, too, the doctors might have noticed something. Motor difficulties, compulsive behavior—but they said nothing.

Anat wouldn't be deterred. She knew it from the beginning. His eyes!

She visited countless doctors with Kai. The recurring question: What brings you to me? The recurring answer: There's something wrong with his eyes. The doctors shone their flashlights into Kai's eyeballs and studied them with microscopic glasses. They measured the distance between his pupils, the pressure on his eyeballs. They used drops and mirrors, but in the end, they would always say, "Nothing's wrong." One doctor had Kai undergo an MRI, sedating him so he would lie quietly. The verdict: All good. Autism reveals itself in the eyes—affected babies often look only out of the corner of their eyes, as their pupils respond more slowly to light—but the doctor hadn't figured that out. When Anat insisted, the doctor studied her. "Maybe your son isn't the one with the problem. Maybe it's you."

"What do you know?" Anat huffed. "You've got your books. I have my motherly instincts. I know my child better than anyone."

In retrospect, that's the big lesson she drew from that time: "Other mothers often ask me what to do. The first thing I tell them is, *Trust your intuition.* You probably know better, even if you don't have the words to express it. Young parents, especially, hesitate to trust their feelings. They do research online, in books. And if a doctor tells them everything is fine, they believe it, because a doctor is supposed to know. But that isn't always the case. Parents have their own perspective, and it's important. I tell them: listen to your feelings, until you find a doctor who really knows better."

Anat gave up on regular doctors. She brought Kai to a world-famous practitioner of Chinese medicine who was visiting the Weizmann Institute. "Well-observed," the woman said. "Something is wrong." Anat's heart skipped a beat. "I can't do anything for Kai," the doctor said. "But I know a colleague in China, a real master. He specializes in patients like Kai. He sticks little needles into the patient that have to stay there for hours."

Anat felt a surge of hope. The professor wanted to try inserting a few needles herself. Kai fidgeted, couldn't sit still. "Is he always like this?" the doctor asked.

"Yes," Anat admitted. "After two days in kindergarten, they wanted to kick him out because they needed one teacher just to keep up with him."

"If that's really the case," said the doctor, "even the master can't do anything for him."

Kai's urge to move, his constant running, jumping, climbing, was overwhelming. He couldn't remain on task. There wasn't much that could hold his interest. He built LEGO towers, looked for buttons on electric tools. Technology fascinated him, in particular the vacuum cleaner, but even that couldn't keep his attention for long. He was constantly in motion, en route to look at something, fetch something, hug someone, talk to the person, run away again. It was so obvious that something was wrong. Even the neighbors asked if they had taken him to see a psychologist.

* * *

It took them awhile to get that coveted appointment with one of Israel's best-known child psychologists. Henry already had a diagnosis in mind but didn't want to intervene in the process. The best doctor should look at Kai. He would identify what was going on. Soon Kai would be talking like the other children, learning patience, finding calm.

The psychologist sat before them with a look that reflected his sense of importance. Anat told him she had had mixed experiences with doctors. The doctor raised his eyebrows and took a book off the shelf. "I have written a couple of books, standard works. You can find them in every medical student's room." He inscribed it and handed it to Anat. "How can I help you?"

Language delay, huh? Well, well.

Can't sit still? Well, well.

Runs up to strangers and hugs them? Well, well.

Loves technology? Well, well.

Touches everything that lights up? Well, well.

Who have you seen so far?

Oh, a bioenergetic therapist? Well, well. And he said that Kai recognizes complex geometric structures? Well . . .

He studied Kai. "He actually seems a bit stunted to me. Younger than his age. Anyway, let's give it a shot."

He sat Kai down at a table and laid out pictures before him. Kai had invented a new game over the past few weeks. He would stare at adults long and hard. That made them all jittery.

The doctor showed him one picture after another.

Point to red please.

Point to the circle.

Point to the carrot.

How many corners does the hat have?

Please place the triangle on the picture of the triangle.

Kai returned to his favorite pastime. He took the triangle, settled his gaze on the doctor, and just waited. He waited for a reaction, a surprised look, an impatient gesture, an admonishing sentence. But the doctor didn't do him that favor. He didn't say or do anything. If he glanced over at Kai, it was only cursorily. Not getting any reaction, Kai turned the game up a notch. He moved the triangle very slowly and placed it intentionally in the wrong place, in the hope of eliciting a furrowed brow, a reprimand, anything whatsoever. Kai was focusing on the doctor rather than the task. But the old man hardly looked up from his notepad. He monitored Kai's activity only from the corner of his eye, over his glasses. The only thing that interested him was whether Kai placed one triangle on another, and how long it took him.

So, you think you can evaluate Kai? Anat thought. *He's evaluating you.*

"The matter is simple," the doctor said, after inviting Henry and Anat into his meeting room. "Kai has attention deficit disorder. You know what that is?"

"Yes," Henry said. ADHD. The brain's frontal lobe has trouble filtering out unimportant stimuli. The child is overwhelmed, and has trouble staying on task.

This is what Henry had suspected. He had already researched how people feel when they suffer from ADHD. They never find any peace of mind. They feel like banging their head against a wall. They can't help but run and jump around.

"I've also concluded that Kai's development is stunted," the doctor declared. "His language capabilities, his motor functions, but also his intellectual and emotional development are a good year behind. He is functioning on the level of a three-year-old."

"And what do you suggest we do?" Henry asked.

"We are talking about a mild case, so medical treatment won't be necessary."

No Ritalin, then. Henry was relieved to hear this. Ritalin can certainly help, but it can be bad for a growing body.

"I recommend taking Kai out of kindergarten and sending him to a preschool for special needs children."

* * *

"Will I have to change kindergartens?" Kai asked on the drive home.

"No," Henry said.

A special needs school could be helpful, but Kai would feel stigmatized. Attending a special needs school makes it hard to ever return into the "normal" system. Anyway, mild ADHD wasn't such a big deal. Every class has someone with ADHD in it. Didn't

Albert Einstein have it, too? Henry himself was a bit fidgety, compulsive.

"Kai got a lot from me," Henry says. "When I wake up, I am instantly at top speed. Our girls woke up very slowly by comparison. I'm instantly *on*. Kai was like that too. He wasn't afraid of anything. When I said, 'Kai, let's do this or that,' he would jump up, instantly on fire—he had this incredible energy. I was happy to finally have someone who said, 'Yes, let's do this. Let's conquer everything.'"

Henry laid out his parental strategy accordingly. If Kai was bursting with curiosity, needed to move, then they wouldn't slow him down—they would encourage him. They would show him what he wanted to see, teach him what he wanted to know. Take off, conquer, discover the world: It would be so wonderful. He had no idea how much this would harm Kai.

5

The Suspicion

No one who meets an autistic child
is unchanged by the experience.

In time, it became harder for strangers to reciprocate Kai's love. This was painful because Kai still loved them. He still ran over to neighbors, to pedestrians, to the old folks on the benches. He didn't hug them anymore. He had learned that not everyone liked that. He talked to them, as his language teacher had taught him, syllable by syllable: "Hello. I'm Kai."

He was five years old now. His curls were even curlier, his eyes bigger, his nose even stubbier. On sight, strangers still greeted Kai with a smile. But his words didn't have the same magic as his hugs. The more he said, the more people turned away from him. Henry and Anat saw it coming, the ungodly moment when the look on people's faces started to change. Kai noticed it too, but he didn't let that hold him back. He wanted to keep his friends close and ended up scaring them away altogether. He stayed behind, alone, angry at himself and the world. This hurt Henry and Anat terribly.

It wasn't hard to recognize the flaw; it wasn't hard to see why people looked at Kai differently now. Something about him screamed, "*Me, Me, Me!*" Kai's world, which had once revolved around other people, now revolved around himself. The charm of his idiosyncrasies was wearing off. The kind of strangers who had once praised him as exceptional now whispered, "What a strange kid." They didn't know what to make of him.

Henry and Anat had hoped that preschool would do Kai good: new friends, experienced teachers, fresh discoveries. They enrolled him at a so-called open school, a democratic school if you will. Each child could decide for themselves whether they wanted to spend time learning or playing, whether they wanted to remain in kindergarten or actually go to school, whether they wanted to play in the sandbox or discover the wonders of numbers and letters. Every child could learn at their own pace.

Henry and Anat started to notice the way the other kids were developing. They left the sandbox for the blackboard; toddlers became schoolchildren. Only Kai stayed behind, still the little boy who followed airplanes in the sky, who only wanted to cuddle and listen to lovely fairy tales. The teachers started treating him differently, charitably, like he was slow. Kai stayed in the sandbox.

The other kids still liked him. But even if they showed up for his birthday, Henry and Anat realized they were no longer really his friends. Indeed, they couldn't be. Kai didn't see this, but he felt it. He courted them more intensely, offering them toys, talking at them. But it was no use. Eventually, the tears started to run, and he withdrew to his room, stranded between two worlds of friendship, one jam-packed with everyone who had ever greeted him, the other gapingly empty. He had the most and the least friends. He was the loneliest boy anywhere.

One day, Kai came home beaming. Anat was pleased to see this: "What's going on with you?"

"Me and Jacob threw stones," Kai said.

"Stones?"

"Yes, at cars."

Kai didn't know what he was doing, but his friend certainly did. A few of Kai's classmates had started taking advantage of him. They saw that the teachers treated Kai differently, that they let him get away with everything. They got Kai to do their bidding.

Who threw the stink bomb? Kai.

Who broke the window? Kai.

Who threw the fireworks at that group of kids? Kai!

The fireworks caused a huge outcry. Kai never could have thought of that by himself. He was afraid of fire and loud bangs. He couldn't stand noise.

What would become of him? With every inch Kai grew, their worries grew also. He started throwing tantrums. For no apparent reason, he would hurl himself to the ground, shrieking, flailing, pounding his fists. They had stopped taking him to the movies a while ago. On public transport, you couldn't stop him from talking to every single person, from the sleeping drunk in the back all the way to the bus driver.

Kai was also becoming very particular about food. There were few things left that didn't offend his increasingly refined nose and palate. He still ate white rice; meat that had been carefully rinsed of all sauce; sandwiches with peanut butter and cottage cheese; and cornflakes, unless Mom had poured a new brand of milk on them

("It tastes different"). In preschool, it took months for Kai to try the cafeteria food and sample a few macaroni.

This was supposed to be ADHD?

Henry kept noticing other tics. Kai took everything literally. If his sister said, "My ears are burning," Kai looked at her with wide eyes and screamed: "That's not true! You're lying!" This quirk often led to disagreements. Other children often said things they didn't mean. Kai took this to heart: his friends were lying to him!

Kai developed an almost manic love of technology. He immersed himself in computer games, constructed LEGO towers that deserved architectural prizes, and completed jigsaw puzzles so quickly it took Henry's breath away. He didn't even consult the picture on the box. He just looked at the shape of the tiles. If Henry was in his lab until late and called Kai to wish him good night, his son only talked to him about his current puzzle. What's more, he spoke as if his father were in the room and could see the puzzle too. Kai didn't seem to get that they were speaking over the phone. That's almost autistic, Henry thought. He couldn't be . . . ?

* * *

The World Health Organization classifies autism as a disease, a developmental disorder. Its cause, Henry had learned in college, is unknown. There are more than sixty theories seeking to explain the origins of autism. It took until the end of the twentieth century for a consensus to emerge. Certain people are genetically predisposed to autism. It is likely triggered by environmental factors, such as alcohol, mercury, or medications, when the brain is developing in utero. There must, however, be another factor. Why else would there be cases of identical twins where one child is autistic and the other is not, though they share the same genes and shared the same womb? It follows that what happens in the years after birth can also be decisive.

According to most influential books on the subject, autistic people are not social creatures. They cannot empathize with other people. They aren't even particularly interested in them. They are reclusive and avoid eye contact. Many of them have compulsions. They place objects in a row, repeat sentences, or constantly rock back and forth. They hate change.

Autism manifests in various ways. Experts speak of a spectrum. If you know an autistic person, you know one and not all of them. Each person is different. Some require foster care. Others are superstars of music or math but are incapable of going shopping by themselves. Some live independent lives and resist being labeled sick or disturbed. To them, autism is a characteristic, a feature, like dyslexia or being left-handed.

Asperger's is particularly well-known. It is considered a mild form of autism. People with Asperger's often find a place in the world. Some have so-called island talents, also known as "savantism." If anything, Henry thought, Kai has Asperger's.

There were telltale signs.

His difficulty with language.

The way he talked past people.

His inability to understand figures of speech.

His clumsiness.

His tantrums.

The fact that he felt less pain than others.

Autistic people have a counterintuitive pain threshold. They can be seriously injured and not complain at all, but they can also shriek with pain at the slightest touch.

Ever since psychologists Leo Kanner and Hans Asperger first wrote about autistic children in 1943, their fate has touched people's hearts. It is as if these children were from another planet. They seem to lack that essential human quality, the ability to socialize. Their eyes stray into the distance. They hardly speak, and when they do, their expression remains static. It's as if they lack empathy, have no desire for closeness. They don't react when their parents smile, don't lift their arms when their mother reaches in for a hug. No one who meets an autistic child is unchanged by the experience.

As much as the fate of these children has moved people, they have been stigmatized even more. In the Third Reich, autistic people were sterilized or killed. After the war, society locked them away and blamed their mothers for their condition. Until well into the 1960s, it was widely assumed that motherly cold-heartedness caused this form of emotional withering. This theory spawned a medical term: refrigerator mother. Henry contemplated how this must have compounded the pain of these women. Preoccupied by his suspicion about Kai, he had pulled all his old textbooks off the shelf. He soon returned them, relieved of his worry. Kai couldn't possibly be autistic.

Rituals are an essential characteristic of autistic people: that famous tendency to organize things, to constantly repeat the same actions. Kai didn't do that.

Autistic people were also known to avoid eye contact. Kai looked you straight in the eye. Autistic people were reclusive, didn't approach people. That couldn't be said about Kai. Henry didn't know anyone who sought social contact as much as his son. Psychologists call this being "hypersocial." A hypersocial autistic person? That was unthinkable.

No, that wasn't it. But what else could it be? Henry decided to look into it himself. To that end, he would take a yearlong sabbatical in the United States. There was so much he wanted to know. What exactly is ADHD? What is autism? How can one help the children affected by them? What does science know about these conditions, and how is this knowledge being implemented in medical practice? First and foremost: What could he do for Kai?

6
San Francisco

Henry had been busy writing letters.
He would meet famous colleagues, legendary scientists.

Henry was geeking out. California awaited him, that golden land of research! Stanford, Berkeley: some of the best research universities in the world. The University of California alone had brought forth ninety-one Nobel Prize winners. Albert Einstein had taught there. Silicon Valley, San Francisco—the concentrated genius and knowledge of a whole generation of researchers. As the trip drew closer, his colleagues' envy became palpable. Henry felt a sense of joy that he hadn't for a long time. The year 1999 was winding down; a new millennium beckoned. Why not dream big? Two thousand would be their year.

He kept busy writing letters and e-mails. He wanted to meet famous colleagues, legendary scientists like Eberhard Fetz, Barry Sterman, or Michael Merzenich. They responded warmly to his overtures. Of course, their respected colleague from Israel could visit them. What a treat. They could discuss the latest research and share their own insights into autism and ADHD. Naturally, they would put him in touch with the relevant hospitals. They found it commendable that their colleague was taking a break from his research, leaving the lab to go poke his nose into the clinics. Everyone should do that once in a while. If only they had the time. *It's not that you don't have the time*, Henry thought, *you lack a reason.* He would tell them about Kai.

First, he needed to attend to his family. He wanted to spend more time with his daughters, Linoy and Kali. Henry felt guilty: Kai required so much attention that they always seemed to come up short. Henry was proud of his girls, who seemed to be sailing through life so effortlessly. They loved Kai, took him along to meet their friends. They knew how to calm down their little brother. Kali needed only to hold on to his hand to mollify him. The girls had the same wild locks as Kai, the same dark eyes and cute nose—the same family face. They loved that their parents weren't sticks in the mud, that they had sent them to forest kindergartens and Montessori schools, that they spent weekends by the sea or up in the mountains, that they flew around the world, almost like riding the bus to them. Oh, and their favorite destination was America, their voices chattering in the backseat as they drove into San Francisco. Beautiful weeks lay ahead, the Pacific, the Golden Gate Bridge, Alcatraz, cable cars, all of it so different from Europe and Israel, urban and rural at the same time, glass skyscrapers here, wooden houses there, something new and exciting waited on the other side of every hill. They ate elderberry ice cream, laughed about dog salons, explored the botanical garden, inhaled the scents, studied the gigantic trees, and went to the nearby beach as often as possible, where the air was thick with salt, the beach was green with seaweed, and the Golden Gate Bridge shone on the horizon. Across the bridge you could get the best burgers. Alcatraz was the only place they hadn't been yet, because Kai refused to go.

Henry enrolled him in a Jewish preschool. The other kids were much younger than he. Kai walked over and hugged them, but most of them didn't like it and pushed him away. The counselors took note of him. He was sensitive: if another child was crying, he cried too. But then he walked over and shoved them. No one could explain that. Two days in, they said that, while they had grown very fond of him—rarely had a child moved them as much—he simply wasn't mature enough for preschool.

Henry and Anat enrolled him in a special needs school temporarily. Soon he would be six and old enough for a real school. The classes were small, with six students in each, and the teachers were well trained. Kai is a loner, they said. He wanders around instead of playing with the others. He can't stay focused on one task. But he'll manage.

One day, the phone rang. It was the principal. Kai had tried to kick him.

Henry and Anat consulted another doctor, a psychiatrist, who spoke to Kai, took a moment to consider, and prescribed him Ritalin. Henry was open to this. It worked for most children, after all. Its effect is paradoxical. Ritalin releases dopamine and noradrenaline. It dopes up the brain, like a mild form of cocaine. If a healthy person takes it, they have trouble sleeping. Why then would it help calm a restless child? Henry asked the same question in San Francisco, Los Angeles, and San Diego. He asked in laboratories and hospitals. No one could answer it conclusively. Many of them said that it wasn't so important anyway. Shouldn't you be happy if a medication works?

Henry had his own opinion on the matter. Was the problem, perhaps, that they were considering disturbances of the brain in isolation? Neurologists think differently than orthopedists, who may look for the source of a patient's back pain in a crooked pelvis or infected tooth. We simply know too little about the brain to treat it holistically.

Kai took the pills without grumbling. Now everything would be better. Ritalin would give him some peace of mind. He'd be able to focus, keep his feet still, sit down, draw, do crafts, and he would get along better with the other kids. Kai would be happy, the family's suffering would finally be over, and Henry could continue his research to understand Ritalin and find a way to make it redundant. That was the plan.

Unfortunately, Ritalin made everything worse. Kai stopped sleeping. His head ached terribly. He couldn't focus at all anymore, couldn't compose himself. His brain was totally unbalanced. Kai stopped eating, and he screamed, swore, and spat more than ever before. They took him off it.

Henry and Anat sat despondently in their beach house. They had to get out. Take a family vacation.

7
The Cobra

"Kai! Kai!" Henry shrieked.
Kai couldn't hear him.

Where on earth was he? Henry looked around: a few mud huts, a dog dozing in the shade, the streets shimmering. No sign of Kai. They were standing in a village in India in the middle of nowhere. This couldn't be happening.

Two months earlier, they had packed their backpacks and taken off. A trip through Asia: leaving behind the hospital, the lab, just letting it all go for a while. The next few weeks would be devoted to family. Linoy, Kali, and Kai should explore, observe, marvel, and grow in the process; learn to accept that every culture has its own rules, its own music, its own food. The children were well-traveled enough by now to encounter Asia, the most exuberant continent.

This trip, Henry thought, would be especially important for Kai. His eyes would be sated, his mind filled with new impressions, and he could walk off his antsiness. Kai would find peace.

They started in Thailand. Koh Phangan, Koh Samui: snorkeling in the green-blue sea, eating coconut for breakfast, drinking mango lassis, eating pineapple on the beach amid the passing vendors. The word *massaaaage* resounded every half hour or so. Henry flagged down the men and women with the pointy hats. Everyone got a massage except Kai, who didn't like the feeling of oily sand on his skin. At night, they explored the local market, sampling fruit they had never seen before—the spikey rambutan, the meaty durian. Back at home, they ate fish fresh off the grill, except Kai, who couldn't stand the smell. They visited temples, wandered through the jungle. They even rode elephants. How prickly the skin felt through their pants! How they swayed as the elephants walked. The whole jungle seemed to dance.

They stayed longer in one hotel, dreaming the days away. Kai was quiet and relaxed for the first time in years. It was beyond lovely. On the day of their departure, the entire staff lined up in the lobby, and they each hugged Kai and waved goodbye. The whole family was surprised. They learned that Kai had been visiting with the employees every day, every last one of them, from reception to the kitchen, from the gardeners to the maids. He had struck up conversations and learned all their names, helping them work and telling them stories. They loved him. It was almost like back when Kai was the most popular kid in the neighborhood. Their little Kai! He waved back, beaming.

They traveled on to India, toward Dharamshala. Henry explained to the children that they would now be learning something about Buddhism and meditation. It came off more heavy-handed than he had intended. He wanted to show the children something he had learned in the lab: that everyone saw the world differently. They moved at a leisurely pace, often lingering for days in one town, taking part in local life. Henry studied Tibetan medicine. The girls took painting classes, drawing so-called *thangkas* on linen, meditation pictures they had seen hanging in temples. In a monastery, they all learned to meditate, sitting in the lotus pose among the monks. Kai was even calmer. He had taken a liking to it all. With big eyes and even bigger questions, he approached the foreign reality.

They hired a driver to take them to Nepal, the ceiling of the world. The way there alone took their breath away. The meandering roads! Outside the jeep, the mountains towered, and the valleys gaped. They could hardly bear to look. Suddenly, the car stopped. A landslide. "It will take days before it's cleared up," the driver said, eventually.

What to do? Turn around or dig in? A group of mountain farmers, who were standing around inspecting the damage, pointed them to a hostel on the other side of the river. The bridge that led over it, hanging three hundred feet above the water, reminded Henry of Indiana Jones movies. Today, twenty years later, he would never set foot on that bridge, but back then . . .

Off they went, spurred on by the belief that the risk would strengthen Kai. And he was eager. He stepped onto the bridge, carefully at first, one foot after the other, until Henry lifted him up onto his shoulders—better safe than sorry. Kai now towered above the banister rope, and Henry worried he might topple off into the abyss. He set him down again, placed Kai's left hand on the rope, took his other hand, and they fumbled their way across, hearts in their boots. What on earth had possessed them? It took two hours to reach the other side, and in those 120 minutes Henry's hands cramped, the sweat ran down his back, his breathing became shallow, and he was never far from losing consciousness. Kai squeaked along happily. Finally, they arrived at the hostel. They ate, slept, and lingered for a few days until the road had been cleared, and then they had to walk the whole long way back.

Next stop: Kathmandu. Kai, who seemed to cast off all fears, kept running away. Only the meat market made him shudder. It smelled so strong that they all held their breath and dashed to the other side. They emerged vegetarians; none of them could eat meat for the next two years.

The Himalayas loomed over them. They went for long hikes. On one such trail—not as dangerous as the bridge but still pretty steep—Kai took off down the slope. Henry jumped after him, a 45-degree incline, and both of them slid down hundreds of feet of gravel, through thorn bushes, finally plopping in a stream in the valley. They returned wet and scratched up, though in truth it was only Henry who was scratched. Kai, miraculously, was unscathed. He apparently thought it was funny. "What was that?" Henry snarled at him. He didn't know this was a mere prelude.

They spent a few starry nights at the foot of Mount Everest. They experienced a thunderstorm like they'd never seen before and were so transfixed by the lightning and great balls of fire that they spent half of the drive back to Dharamshala recalling the details. They arrived tanned, singing, so accustomed to dawdling that they made a stop in a hamlet that had nothing to offer but vegetable curry. Sated and satisfied, they took a stroll through the village.

Where was Kai? Henry looked around again. He saw the same mud huts, the same dog in the shade, the same shimmering streets, and still no trace of Kai. Henry looked at Anat and shrugged.

How often had he disappeared on them already? They had lost him five times in Kathmandu alone.

The first time it happened, their hearts almost stopped. They shrieked his name, the whole family searching frantically amid the bustling crowd, only for him to suddenly reappear with that look in his eyes that made it so hard to be mad. "I was over there," he said. "They have prickly fruit." They scolded him. "Think of what could happen! Your sisters don't run away like that." Kai appeared contrite but then disappeared again soon after. They took one look at the map, and already Kai had taken off down an alleyway because he heard some music. They were chatting with a local, and already Kai was in hot pursuit of a dog or a water fountain. When he disappeared at a festival, they climbed a nearby hill in the hope of spying his T-shirt in the crowd. Suddenly, he was standing right next to them. Henry started to grasp that they were seeing things the wrong way around: Kai hadn't lost them; they had lost Kai. In fact, Kai was the responsible party. At least, he always seemed to know where they were.

But where was he now? Perhaps over there, on the square, in the crowd? It looked like the whole village was assembled. Kai was probably waiting for them there.

From far away, they heard tooting and buzzing. A musician, maybe? Henry, Anat, and the girls approached a wall of backs. They tried to see what the crowd was looking at. Henry stood on his tiptoes, and the children bent down to see through the thicket of legs, but despite their peeking over and under, they couldn't see anything. They pushed in farther. "There!" Kali said. She could finally see what was going on—a snake charmer had cast his spell on the crowd. The old man was sitting with his legs folded, a dirty orange turban on his head, a pungi flute at his lips, which sounded a bit like a bagpipe. A cobra danced in front of him, a real one with poisonous fangs, meandering back and forth in a threatening trance.

How this unexpected act excited them! Anat got out the video camera. The crowd was rippling to the rhythm, but then suddenly fell completely silent. A little boy had emerged from their midst and was walking toward the cobra.

"Kai!" Kali shrieked.

The snake charmer's eyes widened, the village held its breath, the flute kept playing, the snake kept dancing, and Kai walked over to the cobra, all the way up to it. Slowly he lifted his little hand, leaned in, and started petting the snake. Tap, tap, tap. "Kai!" Kali, Henry, Linoy, and Anat shrieked. The snake charmer's eyes were now so wide with alarm that he stopped playing. The audience stood stock-still. Kai kept petting the snake, mischievously tapping its head, staring at it with the same expectant look he had shown the Israeli psychologist. *Let's just see what happens.*

"Kai, Kai, come here!" Henry shrieked. The whole crowd started calling him, summoning him back. The snake charmer, who didn't know what to do, put the pungi back to his lips and kept playing; he was playing for Kai's life. While the snake threatened, swinging, hissing, and Kai tried to get its attention, Henry squeezed his way through the crowd, but nobody wanted to budge—this specter was too wild to look away. Finally emerging at the other end, Henry grabbed Kai by the shoulders and pulled him away.

"Kai!" he said quietly and hugged him tight.

"I was just . . ."

Anat and Henry spent the rest of the day in shock. What if the cobra had bitten him? A young child poisoned in the hinterlands, hours from the nearest hospital. They had been so lucky.

Their driver later explained to them that the snake must have just woken up from a long nap; it must have been terribly confused by all the light and movement. Cobras are deaf, so it couldn't even hear the flute. Petting it would have been just one among many distractions, less threatening than the general glare and hovering instrument.

Henry and Anat lay awake that night. Things couldn't go on this way.

8
The Fox

Distress and cunning were indistinguishable from one another.
It almost reminded them of Kai.

Lausanne, Switzerland: a restaurant close to the International Olympic Committee's headquarters. A young woman and a man in his mid-fifties are eating lunch. They are father and daughter. One can discern this from their eyes and familiarity. A quick espresso to top it off. The waiters watch impatiently, sternly rattle the cutlery. It's almost 2:30 p.m. The two have gotten carried away chatting. They are talking about the past, about Kai.

HENRY: Remember the cobra?

LINOY: Of course, I do. We even have it on tape. We were so shocked! I watched it again recently when I cut the video for Kali's wedding. Someone always had to hold Kai's hand to make sure he didn't run away. On the old videos, it's often Kali who grabbed hold of him.

HENRY: Yes, she was his guardian. More bodyguard than mother.

LINOY: And you guys were always filming. Zoom in, zoom out.

Henry: I have no talent for that.

Linoy: True. Sometimes you can't even see us.

Henry: Do we have videos of us in Heidelberg?

Linoy: More from the US.

Henry: We had a lot of fun in Heidelberg.

Linoy: I don't really remember.

Henry, *hesitantly*: You were five, a little girl. . . . One thing I meant to ask you: it must have been challenging growing up with a brother who could be so difficult, who demanded so much attention.

Linoy: I took him along to the mall once and, I don't recall why, but he threw a fit. I started crying. And I remember he stopped and looked at me, like, "What's going on? What have I done?"

Henry: When did you first realize he was different?

Linoy: I don't remember ever thinking like that. He was just a bit wild. He always attended our birthday parties and made friends with our friends.

Henry: How did they respond to Kai?

Linoy: They always ask me how he's doing when I run into them. Everyone who met him wants to know. He leaves a big impression.

Henry, *hopeful*: So, you guys didn't even notice that he was drawing all that attention to himself?

Linoy: Kali did. She has said that to me. I don't remember feeling that way. I just remember thinking sometimes that he was a bit out of control. And that I was scared because we always lost him. Aside from that, I think, it was actually quite relaxed, not particularly chaotic. I started seeing things a bit differently when I was a teenager. Before that, I just thought of him as a baby who had to be taken care of.

Adults can learn a lot from trying to see the world through a child's eyes. Just as children aren't particularly concerned about skin color, they at first don't pay it much mind if a child behaves a little differently. That's just the way it is. Kali and Linoy didn't even see Kai as different. Sure, he had some quirks. Linoy will never forget how he petted the cobra, or how he jumped out of the boat in Thailand and she grabbed him by his bathing suit. You never knew what you were going to get with Kai. But they didn't find him different, no. He was just Kai.

They weren't even particularly considerate of him. They took him everywhere and treated him like any sibling. They fought, pulled each other's hair, made up again, laughed, played, cuddled, and always found it curious that their parents were so protective of their brother, that he could get away with so much. Then again, how could Henry have been strict with Kai? He loved his kids and had promised them at birth to always be gentle and patient. He knew what it meant to be raised by the whip. Growing up, he was paralyzed with fear when his grandfather got mad; his chest tightened up, and he felt a tingling sensation inside, like his veins were full of ants. Henry never wanted his children to feel that way. They should take life easy, learn to fly.

What was he to do now that one of his children was threatening to fly away? How strict can you be with problem children? How much freedom can you grant them?

* * *

They returned to San Francisco with the terror from the cobra episode still pulsing through them. Their worried eyes followed Kai's every move. And yet, they still granted him liberties that Kali and Linoy could only dream of. They didn't force him to eat his salad or do his chores, didn't reprimand him when he was rude. Observing Kai, Henry often found it hard to distinguish between distress and defiance. There was no distinction anymore between "poor Kai" and "naughty Kai." Sometimes he had the impression Kai knew all too well how to play his cards, that he was a master of manipulation.

Henry remained lenient, usually tried speaking with Kai. He enrolled him in a Montessori school for their last few months in the United States. The Montessori system promised structure as well as freedom. Parents with difficult children are often drawn to this type of school. It was founded by Maria Montessori, one of the first women to study medicine and get her doctorate in medicine. She was brave and headstrong, and she loved children. After completing her own education, she took a job in the psychiatric wing of a hospital, working with disabled children, lost souls whom she hoped to bring out of the shadows. She invented games for them, known as sensorial materials, which stirred their imagination and awoke their curiosity. And the children made progress. It turned out that some of them weren't disabled at all, just stunted and neglected by a society that didn't provide their minds with an opportunity to thrive.

Maria Montessori founded nurseries, where children could learn freely, without reward or punishment. She believed that children wanted to learn and participate in the adult world, but that one shouldn't rush or force them, letting each child decide for itself, deliberately, on its own terms. A conspicuous number of our most influential entrepreneurs are Montessori graduates, such as Amazon's Jeff Bezos and Google's Larry Page and Sergey Brin—little troublemakers, nerds, geeks—who still love to talk about how going to the right school set them on their paths.

Montessori schools also set limits. The students are expected to clean up after themselves and listen to each other. "It is our duty to prevent the child from doing anything which may hurt or offend others," Maria Montessori once said.

Kai and his teachers developed conflicting opinions about what constituted such an offense. Kai was offended by people using metaphors or suddenly changing plans, while the teachers were offended by the tantrums he threw in response. When he spat at a classmate—he spat a lot—that was it. The child's parents considered suing Kai; you can do that in America. Kai got kicked out of school. Their year in the United States was coming to an end, and they would just have to keep trying in Israel.

* * *

During those last few weeks in San Francisco, Henry settled more conclusively on autism. The cobra incident had been pivotal. So had Michael Merzenich. Henry met with his famous Berkeley colleague whenever possible. Merzenich had once been considered a crank. He had questioned the dominant consensus, which stated that the human brain was immutable after childhood, that it couldn't change, grow new connections, or develop new abilities. Merzenich went so far as to say that we could change our brain through thought alone, that we could cause it to grow and make certain diseases disappear. A scandalous claim. It contradicted a natural law: a window can't open itself; it requires a determined hand or a gust of wind. A ball can't choose to fly, a foot has to kick it. Only something material could bring about physical change, as determined by the law of causal closure. To many of his colleagues, Merzenich's claim was tantamount to saying that you could open a window or move a ball with your thoughts alone.

Today, Merzenich is an icon. The old experts turned out to be wrong, their assumptions false. Today, we know that the brains of taxi drivers grow when they learn their city's map by heart. We know that people can learn new things well into their old age, and that particular thinking techniques can alleviate schizophrenia and depression. We know that sea nomads in Thailand who live on boats learn to read underwater, meaning that their eye reflex has changed in a way deemed impossible. The brain forms itself. There is a word for that: *neuroplasticity*.

But it wasn't this insight that drew Henry to Merzenich. He wanted to know more about another of his colleague's theories. Merzenich had studied autism and determined that it was caused by neurons, the cells that transport signals—Henry's field of expertise. Some neurons amplify signals, ensuring, for example, that the command to take your hand off a hot stove arrives very quickly in your brain. Then there are neurons that inhibit signals, say, the urge to pet a cobra. That's what Henry wanted to speak with Merzenich about.

"Autism? Your son?" Merzenich looked concerned. Over the next few days, he let Henry in on his theory. After they had discussed everything, he turned to him and said: "Why don't you visit Lynda and Michael Thomson in Canada? They specialize in both autism and ADHD. Excellent scientists. Most importantly, they aren't armchair scientists. They apply their research."

* * *

Henry devoted his last days in America to the family. They went to the beach often. The water was even colder, the wind still blowing briskly, the seaweed on the dunes getting musty. Those were pleasant days. Henry and Anat watched Kai hopefully. He was bronzed by the sun and laughing a lot. He wrestled with his father. He had gotten so strong!

They decided to go on one final road trip. They drove through Arizona, the kids squabbling in the back because someone refused to comply with Kai's desired seating arrangement. They drove along a lonely, endless road, white stripes at the shoulder, yellow stripes in the middle, red mountains on the horizon. Suddenly, in the distance, something appeared in the middle of the asphalt. Henry slowed down as they drew closer.

"A fox!" Henry exclaimed.

It stood there pathetically, panting, with its red bushy tail and head lowered, its tongue hanging out. They drove around it slowly.

"It's thirsty. It's going to die," said Linoy.

"It's hungry," said Kali.

"It's going to die," said Kai.

They pulled over sixty feet down the road. The fox was hardly moving now. It turned its head ever so slightly, looked at them, pleading. "Oh, my god, we have to give it some water," said Kali.

"And something to eat," Linoy added.

Kai remained silent.

Henry cautiously stepped out of the car and placed a bowl of water and a bit of bread and cheese on the tarmac. He got back in, drove a little way down the road, then pulled over again.

Was the fox strong enough to reach the food? Suddenly, it lifted its head and tail and shuffled toward it. As it ate and drank, something like a smile rose to its lips, the way dogs smile when they're sated and comfortable, sprawling next to a barbecue grill on the lawn.

The Markrams made a U-turn and drove back. They had to see this. As they got closer, the fox let its head, tail, and tongue hang. It looked abject and miserable again, like it was on its last legs. The family laughed and gave the fox a bit more food, and tried to explain it all to Kai, who was staring out the window with his mouth open. And somehow, though he was oblivious to metaphors and tactics, Kai seemed to understand it more readily than one might have expected. The fox also knew how to play the cards it had been dealt. In the following days, Kai talked about the fox more or less nonstop, how it had gotten the food and water it needed to survive in the desert by pretending to be sick. What a smart fox! Its trick had worked so well! Distress and cunning were indistinguishable from one another. The fox was a master of manipulation. Somehow this seemed all too familiar.

9
Lynda Sees It First

Lynda loved dogs and koalas.
She didn't have to say much to win Kai's heart.
Then she looked into his head.

"Point to the pencil," Lynda said.

Kai looked at a picture of a nail and a pencil and pointed to the pencil. Lynda noted this.

"Please, point to the red dot, Kai."

Kai had a page in front of him with four dots: blue, yellow, red, and light blue. Kai pointed to the red dot, and Lynda noted his response.

More than any child, Kai had awakened Lynda's protective instincts. He didn't seem as if he was almost six years old. He seemed more like three. Everything he said or did was like a plea for love and affection. Obediently, he did as she instructed.

Kai liked Lynda. She had the thickest hair in the world, she was nice to him, a picture of a koala hung in her room, and she had a dog. The dog nudged him with its snout. Kai stroked its ears. How soft they were! When Kai got bored, he could walk around in her garden. Then Mom called him. He had tasks to do.

"Kai, please click the mouse when you see the number one. But definitely don't click if you see a two. Do you understand?"

Kai gritted his teeth and stared at the screen with grim determination. When a number appeared, he didn't just click, he scrutinized it carefully. That way, he missed a few but didn't make any mistakes. He never made the mistake of clicking a two, even if it popped up quickly and treacherously between ones. Lynda watched, scribbling notes.

The Markrams had arrived in Toronto the day before. When Henry began taking an interest in ADHD and autism, he inevitably came across Lynda and Michael Thompson, a psychologist and a child psychiatrist, respectively. The two of them ran a private hospital, one of the most hopeful places for ADHD in North America.

Lynda's story was remarkable. Her love of the German language had brought her to Hannover, in Lower Saxony, as a young teacher. She admired one of her students there. He was highly intelligent, had great marks for class participation, but tested miserably. What's more, he was difficult to the point of being out of control. The other teachers wanted to kick him out, but Lynda fought for him. He simply had to get a chance to graduate. She lost that battle, and that's when she decided to become a child psychologist. Her dissertation dealt with Ritalin and its effect on children with attention deficit disorder. She met a man, a psychiatrist named Michael, who himself had traits associated with ADHD. The son they had together developed the disorder. When he hit puberty, he slipped away from them almost entirely. Everything bored him. He was distracted. He had to drop out of public school, enroll in a private one. The only class where he showed any promise was PE. Maybe he'll go pro, the PE teacher said, trying to console them.

How can we help him? Lynda and Michael wondered. They didn't want to give him Ritalin, which is prescribed to three of every four children with ADHD. Lynda knew all about its side effects from her dissertation research. Ritalin can make you drowsy, dizzy, and a whole lot more. And it doesn't cure you. It merely dulls the symptoms, and if you discontinue it, you're as bad off as before.

She heard of a new, unlikely method: neurofeedback. Patients learn to steer their own brain waves. Mainstream doctors considered it quackery. Epileptics and people with ADHD swore by it. She wanted to give it a try. What did they have to lose? Her husband, Michael, took part in a workshop. They wired up his head and sat him down at a computer. A bar appeared on the screen. When Michael strained his attention, the bar rose, and when he got distracted, it sunk. The goal was to balance the bar somewhere in the middle.

After several treatments, Michael felt better. Excited, he called his wife: "There's something to it."

They moved to the United States for a year to learn how to do it themselves—no one in Canada offered that kind of hocus-pocus. Their son learned it too, and his grades improved. Soon he was able to leave private school and transfer back to a "normal" one. He graduated with honors, went to Penn State, and became a well-known sports psychologist. Many studies would go on to prove the efficacy of neurofeedback. Even the American Academy of Pediatrics would eventually acknowledge it as a viable alternative to Ritalin. Henry couldn't have known that this would happen back then, in the year 2000, but he did know that all of today's standard practices had once been considered experimental. He'd always been drawn to professors who thought outside the box.

* * *

In Toronto, the tests continued, and a few days later, Henry and Anat were sitting back at Lynda's desk. "So," she began, "we did all the tests and have arrived at a diagnosis." Henry's pulse rate rose. Anat went pale.

"We did a picture-vocabulary test and found that, linguistically, Kai is indeed a bit behind. Eight months or so. His scores line up with the average results of a child who's four years and nine months old. I'm not worried about that. It may be related to the fact that the test was in English. Something else, though, made me think. Remember the test where Kai had to click a button when he saw a certain number? He hesitated and made no mistakes. That's not how children with ADHD respond to the test. They're impulsive, click intuitively. Contrary to what some colleagues may have said, Kai does not have ADHD.

"In the early years," she continued, "autism is often mistaken for ADHD. In both cases, children have trouble focusing. You can tell the difference when children interact with each other. Autistic children have a harder time with that. They often talk only about their own interests or seem like little professors. They're also very sensitive. They can't stand noise, labels in their clothing, someone combing their hair. Their motor skills are often bad. And they disappear into their own world. Now, people with Asperger's do seek contact with others. They just don't know how to go about it. Children with

Asperger's often speak like adults. And they take everything literally. If you say, I'll be there in a minute, they will hold you to that. When they get mad, they're hard to calm down. They're honest, naive, wonderful people—if you treat them right. The three most important rules are as follows: Be nice. Be nice. Be nice."

"We have developed a method to ascertain if someone has autism," Lynda said finally. "We monitor their brain waves, and if a particular one breaks away, that points to Asperger's. And that's the case with Kai. Other forms of autism could also be at work here. Kai is absolutely a candidate for neurofeedback. The question is whether we can convince him to do it. He has to want it. Ritalin is certainly the wrong way. It makes everything much worse for autistic people."

Henry sat there silently. He heard what she was saying. He didn't hear it as a doctor. He wasn't thinking like a doctor at that moment. He was sitting there as a father—feeling small, vulnerable, at the mercy of Lynda's words—and once she had finished, silence filled the room and his insides.

When Henry and Anat left, the word *autism* followed them, at once silent and shrieking.

* * *

The same diagnosis can elicit profoundly different reactions. Some people are overcome by fear and sadness. They never lose that feeling. They never return to their old life. They are changed. There's one life before the diagnosis and another afterward. If all goes well, and the disease is at least contained, time might bury those feelings deep down, but they never go away.

Others experience the diagnosis as a liberation. Those tend to be the people who have been suffering for longer but didn't know from what. Nothing demoralizes you more than an enemy you can't see. They feel relieved that their adversary has finally shown its face. They can do something about it now. Fight or flee. For a while, they manage to leave the fear and sadness behind them. That's what happened to Henry.

"It felt like our journey had finally come to an end. Now we knew what it was. Sure, in retrospect we didn't really know, but it felt that way. When you hear the statement, 'This is it,' you're so appreciative that someone knows what's going on. You assume they know how to go from there."

They had suffered for so long, each in their own way. Anat protected Kai, showered him with love and understanding; she fled inward, she slowed down. Henry's despair drove him out into the world, into the lab, into libraries; he accelerated. "His father seems agitated, hunted," Lynda noted in her case history. Like all parents with children who suffer, Anat and Henry wished they could bear Kai's burden for him. That would have been easier on them than watching helplessly. Parents are meant to protect their children. Nature arranged it that way.

Lynda had given them hope. "It wasn't a bleak diagnosis," Henry says. The two of them felt as if they had finally arrived, like at the end of a long journey when you arrive at the hotel at night and just fall into bed. You can unpack the suitcases tomorrow.

They decided to give neurofeedback a try. It would be fun for the girls, like those drawing classes in India. Everyone participated, steering bars, dots, cars, and airplanes with the power of their minds, dashing through the overgrown brain maze. Only Kai refused to do it. The helmet hurt his head. He found the gel under the electrodes gross.

His parents' outlook darkened again. Their relief turned out to be temporary. They were far from the end of their journey. It was just the beginning.

II

THE HUNT

1
Powerless

"Most people assumed that I, as a neuroscientist, could help my child more than other parents. They were wrong. I felt even more powerless."

As quietly as autism had sneaked into their lives, it now settled in all the more noisily. Their family life was progressively built around Kai's tics. Of course, he was still the sweet little boy who livened and shook things up at home. He loved skidding in water, so when Anat cleaned the house on Saturdays, she flooded the whole floor, and the kids slid by, laughing, with Kai laughing loudest. On New Year's Eve—usually a quiet evening in Israel, hardly anyone celebrates—they sat in the kitchen, drowsy and bored, until Kai leapt into the room. "It was five minutes to midnight, and he arrived with a bag of confetti," Linoy recalls. "He had spent the whole evening punching it out of paper. 'Everyone get up!' he cried. And we had to count down: ten, nine, eight, seven, six . . . Happy New Year! And he flung the confetti! We really celebrated New Year's. It was wonderful."

Like almost all autistic people, Kai started accumulating more and more rituals. Christmas became his chief ritual. He sorted the carols into his preferred order well in advance. He sang the first word of every song and then the family had to join in. Woe to anyone who started too early or didn't sing along until the last note. The girls sat there with mirthful eyes. Kai was a bigger attraction than Santa.

In everyday life, though, his rituals became a burden. Every morning, it was a struggle to find socks and underpants he was willing to wear. In the evening, he wouldn't go to bed without his cottage cheese sandwich and only if his preferred pillow was in the right place. The battles wore on them. Even Kali could no longer understand her little brother. Henry found it hard to mask his impatience when he was already late for work and Kai was throwing another tantrum over socks.

Apparently just getting started, Kai began enthusiastically throwing his toys out of the window. The neighbors rang the doorbell, and Henry said, "Yes, we know, we're sorry, he's autistic." The neighbors still shook their heads, but at least they didn't ring the doorbell anymore.

Rituals grow out of fears. Kai, the explorer, had become a scaredy-cat. The boy who used to rush over bridges, slide down hillsides, and snowboard faster than anyone in the family, now refused to climb a hill or set foot on a board. Henry's *compañero*, his son, who was just like him—boom! eyes open, ready to take on the world—was no longer available. *No* became the operative word in a life defined by retreat and refusal. Henry had no idea what to do. Was he not being nice and patient, as Lynda had suggested? Was he not doing everything the textbooks said: encouraging the boy, limiting his rituals? It was no use. Helplessly, he watched Kai turn away from the world just as he was turning away from his family. In some ways, they could count themselves lucky. Kai still spoke to them, after all, let them touch him, and even returned their kisses. And yet, it hurt when they would say something nice to him, give him a present, or lovingly cook him dinner and he couldn't muster joy. All thoughts of the future were fraught with fear. Who would take care of him when they were no longer there? Not to mention the constant nagging of their guilty conscience. They didn't have enough time for Linoy and Kali, for friends, for a normal life. In short, they experienced what almost all parents of autistic children go through. Except that, in Henry's case, there was another layer to the suffering.

"Most people thought that, as a neuroscientist, I could help my child more than other fathers. That was false. In fact, I felt even more powerless."

Henry felt that he was failing as a father and a scientist. He couldn't even understand what was happening in his own son's head. All his illustrious essays and prizes were of no use when he was back in Kai's room feeling helpless. He dove deeper into the books, sought more counsel from experts, but it only made matters worse. Other parents could hope. They could entertain the illusion that doctors could help them. Henry knew better. The doctors didn't even know where autism came from or what one could do to fight it. "You're the one who should know," they would say to him. "You're the neuroscientist. We're just doctors."

"You feel guilty. You think you ought to know. It starts to feel like a great big farce. You feel like a failure."

The sabbatical had only made things worse. "The frustration of not understanding is what compelled me to go to America in the first place. That was the mission: to understand. The goal was to bring together scientists and patients. How can we neuroscientists help people in a tangible way? I took a year off for that. I was totally dejected afterward. We scientists had no influence. There was a wall between our research and the patients. We were in an ivory tower, so divorced from the rest of the world."

Henry had to make a decision. He could continue as before, keep applying himself to the minutiae. Some scientists make a big difference with their minor contributions. Many more, however, get bogged down in specificity. A neuroscientist studies memory and delves ever deeper, from the cell to the synapse to the memory synapse. Soon, he will no longer have anything to do with the world outside. The better you are, the more bogged down you get. That's what happened to Henry.

"You join the best lab for memory synapses. You request the best grants available for the study of memory synapses. Then you publish articles in journals that deal exclusively with memory synapses (such journals really exist). You go to conferences for memory synapses, where you meet other experts on memory synapses. And they become your friends, your memory synapse friends, and you love what you do, and you undertake the next great experiment, write the next important paper, give talks, travel the world, collect prizes, until your world is just one gigantic memory synapse."

Kai alone had torn Henry out of his cosmos. He wouldn't be much help to his son as a personified memory synapse. If Henry was honest with himself, his work was no longer in line with what he'd set out to accomplish as a young man. It had nothing in common with the mission his grandfather had underwritten with that check. He had strayed off course. Even if his current path was paved with glory, it remained a dead end.

Henry decided to change direction.

2

Saved from the Wastebasket

Interesting, *the man thought:* a scientist considering autism from a father's perspective.

There they sat: professors from Texas, Kansas, and New York, a tower of papers stacked before them on the table. So many proposals for fewer than twenty-five grants. If they wanted to get through them all, they would have to narrow down the selection.

Let's see what we have.

Cerebellar deformities? Good, foundational research.

Learning disabilities in autism? Crucial, crucial.

Do genes on chromosome 2q cause autism? Perfect, genetic research.

Genes that delay language acquisition? Two important topics in one. We'll gladly support that.

Oh, and here from Israel: a certain Markram, Weizmann Institute. He claims that almost all autism research is focused solely on autistic people's weaknesses, on genetic defects, and on the cerebellum. He wants to study the cerebral cortex, inhibitory neurotransmitters. Our good old friend, Professor Merzenich, has also been researching in that direction.

The professors grinned. If hardly anyone in the field was studying the cerebral cortex, there was good reason for this: because there was nothing worth studying. It's nonsense, some of them said. A sideshow, said the others. The dossier sailed into the wastebasket.

But one person took notice, not one of the power brokers at the table but an unassuming man sitting on the sidelines. He represented the National Alliance for Autism Research (NAAR), a foundation started by a group of parents with autistic children. Nancy Lurie Marks, an heir to the Walmart fortune and mother to an autistic child, was a donor.

Interesting, the man thought. He would tell Nancy about this. She was looking for new ways forward. Unlike scientists, parents don't have time to waste, nor do they always err on the side of prudence. They are willing to take a risk. Once the meeting was over, the secretary plucked the dossier out of the wastebasket.

What the young scientist had to say was impressive. He had reviewed the existing autism medications. All 625 existing patents were geared toward improving neural capacity, toward stimulating the brain. And they were all derived from research that focused on the deficits of autistic people: learning disabilities, speech impediments, genetic defects. Markram questioned that.

And that wasn't his only point of criticism: the existing research was focused on the cerebellum. That, of course, was an important part of the brain but not the decisive one. If a child is born without a cerebellum, it still grows up fairly normally; the brain can delegate the cerebellum's tasks to other departments. The cerebral cortex is much more important, particularly the part that Henry knew best: the neocortex. It contained memory, perception and emotion, those higher cognitive abilities that make us uniquely human—faculties that are impaired in autistic children. That's where Henry wanted to focus his research. He wanted to study inhibitory neurons. At first glance, a sideshow. But didn't people with defective inhibitory neurons experience seizures, just like every third autistic child?

Nancy Lurie Marks liked what she was hearing. That research proposal didn't belong in the wastebasket. This man should get a chance. Her own family foundation would support his research for a year; the second year should be covered by NAAR: $96,800 in total.

Henry was so proud: his first foundation grant.

* * *

Since the early days of medical science, one of the central questions facing practitioners has been how best to introduce their lab discoveries into the real world. How do you determine if a medicine works? Since Paul Ehrlich's day, the solution has been animal testing. This repels many people. "We should . . . refuse to live," Mahatma Gandhi once said, "if the price of living is the torture of sentient animals." What would the pregnant women who were prescribed thalidomide in the 1950s have to say about that? That drug, notably, had not been tested on pregnant animals. How do Parkinson's patients feel about animal testing? We know that their shaking can be alleviated with neural implants only because we tried it on monkeys first.

It's cruel to hurt animals, but it's inhumane to refuse aid to our fellow humans. Humanity has thus come to a tacit agreement that it will continue to conduct research on animals. Henry, who today heads a project that could make a lot of animal testing obsolete, decided to do it also.

His research required breeding autistic animals, autistic lab rats. He administered them substances that were suspected of causing autism: mercury, alcohol, epilepsy medication. But it wasn't quite working. Everything took longer than expected. Under the microscope, he didn't see what he had hoped: that the solution was located in the cerebral cortex and that the inhibitory cells weren't doing their job.

Day after day, he stared at the neurons, delving ever deeper into the cells. He couldn't find anything. Perhaps the professors had been right, after all, and his dossier belonged in the wastebasket.

When Henry drove home at night, where Kai would be waiting for him at the door, he felt lost and infinitely alone.

* * *

It's hard on parents when their children are sick or disabled. Many doctors and psychologists have taken an interest in this burden, writing books and essays about it. Bouma and Schweitzer, Hastings and Johnson, Sanders, Morgan, Weiss, all the usual suspects have weighed in. They all reached the same conclusion: no chronic illness, no disorder or disability causes parents more suffering than autism. Particularly if it's a severe case of autism and the child withdraws from them entirely, offering them no recognizable love, no words, no smiles, their grief is as deep as humanly possible.

The parents cited in these studies all say variations on the same things: It costs you so much strength. You feel wired constantly.

You're always looking out for them. You stop doing things for yourself. You don't go out anymore. You feel ashamed. Friends stop visiting you. You stop spending time with your partner because you're both too tired.

Many marriages break up under that pressure. Other couples pull even closer together. You need each other, depend on each other. This can give a troubled marriage a new lease on life, one that may turn out to be temporary. That turned out to be the case for Henry and Anat.

The two of them met when they were young. Henry, the aspiring neurologist from South Africa, and Anat, the career student from Israel. They were so different, had so very much to say to one another, each exploring a new world. But over the years, they drifted ever further apart. Henry strove onward in the scientific world, going from success to success, while Anat immersed herself in color theory and meditation, searching for the meaning of life. Their marriage wouldn't have lasted as long if it hadn't been for the girls and Kai. But Kai couldn't keep them together forever. One day the two of them realized that they weren't lovers anymore, merely good friends and great parents. They separated but continued to live under the same roof. Their marriage was broken, but their family was not.

3
Kamila, the Astronaut

If not outer space, then at least the brain.

When Kamila was a little girl with blonde pigtails, still too young to go to school, her grandfather interviewed her on tape.

"What do you want to be when you grow up?" he asked.

"I like to travel," said Kamila. "I want to visit a lot of countries."

"Sure, but what do you want to be?"

"An explorer," said Kamila.

But when she finally set off on her first great journey and hugged her grandfather farewell, Kamila felt a deep sadness rising in her. Martial law had just been declared in her native Poland, and her parents emigrated to Germany with her. They ended up in a small village in Lower Saxony, where her father, an engineer, found work on a farm. This was her new home. You'll make new friends in school, her parents said.

She looked forward to it. But she didn't speak German. She felt excluded and alone among her new peers. Her teacher, Mr. Daske, a gentleman with gray sideburns who whacked disrespectful pupils on the fingers with a stick, took up her cause. After school, he flipped through books with her, pointing to the pictures. That's a deer, he said. That's a house. That's a car. Kamila liked him and learned swiftly. Soon she could articulate her dream in German: she wanted to become an explorer, but not one who travels the world. She wanted to be an astronaut.

By then, she was no longer a little girl with pigtails but a young woman with glasses and a serious look in her eyes. Day after day, she read the astronomy dictionary that her parents had given her. She forced her friends to play Mir Space Station. Even watching the *Challenger* explode, the greatest disaster in the history of space travel, didn't discourage her. Kamila held on to her dreams until she reached the age when one needs to start realizing them. Naysayers told her that girls with glasses couldn't go to space—one needed eagle eyes; a teacher admonished her after she returned to school from a year in Mexico, a girl with her attitude would never be an astronaut. Kamila became a space cadet instead; she never studied more than necessary and slacked off in physics and math, preferring to read science fiction books (the Dune trilogy), and watch *Battlestar Galactica* (she had every episode on video).

After graduating, she moved to Berlin to study philosophy, another way of looking at the world from above, another way to answer the big questions. She enrolled at Humboldt University and was surprised to find that, of all things, her required statistics and math courses came easily to her. She just had to learn at her own speed, freely as a student rather than spoon-fed like a schoolgirl. Soon she felt at home again in the natural sciences, just as she had been leafing through astronomy books as a child. She bought physics books, stared at the sky, and when a documentary came on about the International Space Station, Mir's successor mission, she discovered that astronauts could most certainly wear glasses; she couldn't believe she'd let herself be discouraged so easily. She signed up for Philosophy of the Natural Sciences. The class was sparsely attended, but Kamila loved it.

One morning, the professor invited a biologist to the lecture, and the class watched two worlds collide. The scientist sneered in disbelief as he listened to his counterpart's lofty thoughts. "We scientists are different in that regard," he said. "We draw on reality. We conduct experiments, evaluate the results, and develop scientific laws."

"Ah, inductivism," the professor said, now grinning also. "Isn't that naive? People like to think that their view of the world isn't clouded by bias. But that's impossible. Take a scientist from Wichita. Let's say he's trying to establish a physical law about crows. He sees a black crow. He sees another black crow. And then another crow, also black. He formulates a law: all crows are black. But in Africa, there are pied crows, with white chests!"

All this was to say: the world is big and humans are small. We yearn for stability, coherence, and we find it in rules, laws, truths. We devote our lives to acquiring these stabilizers. This age-old desire is sated when we think we've discovered a pattern, when we think we finally understand. This satisfies us so comprehensively that we tend to forget the limitations of our worldview. A person who lives in Wichita doesn't know anything about Africa. The desire for truth distorts the experiment before it even begins. No scientist is unbiased, objective.

"Oh, please," the scientist said.

"What color is this pen?" the professor challenged him.

"It's yellow."

"No, it isn't yellow. You see it as yellow," the professor said. "That's called negative realism."

"Oh, please," the biologist said again, and it went back and forth from there. "I don't care what you think," the biologist snapped. "All I know is that my work makes a difference. It saves lives!"

That debate broadened Kamila's horizons, even changed the course of her life. Intellectually, she admired the philosopher; to this day, she challenges every scientist who thinks he's free and objective to attend such a lecture. But the biologist won her heart. What good was thinking for its own sake, particularly along a line of argument that led only to the conclusion that everything was more or less irrelevant anyway?

"I wanted to do something relevant too. I wanted to do something that made sense," Kamila says. She transitioned to the natural sciences. The great voyage she had envisioned since childhood could finally begin. She wanted to be a biopsychologist, to explore how the firing patterns of neurons affect our behavior. "If not outer space, then at least the brain," she says. "Everything comes from it. It explains everything."

* * *

Kamila sits in her lab in Lausanne. With her cheerful eyes, sharp nose, and round face, she looks a bit like a young magician. She is skinny, her hair still blonde. She switches between German and English, choosing English for more complicated matters, reverting to German when she discusses the past.

Twenty years ago, she was sitting at her small desk in Berlin typing a long letter that started with the words: "Most honorable Professor Singer. My name is Kamila Senderek." That letter was a bit brash. Professor Wolf Singer wasn't the kind of man you just wrote to as a student. He was Germany's best-known neuropsychologist, a member of Academia Europaea and Leopoldina, the German national academy of sciences, as well as the Pontifical Academy of Sciences. If there were a Mount Olympus of the Sciences, he would probably be sitting on it somewhere between Marie Curie and Sigmund Freud, initiating lively debates. Every piece of insight the director of the Max Planck Florida Institute for Neuroscience tossed out in his books and interviews was seized on and chewed over throughout the scientific world.

He certainly wasn't the type of person you asked for an internship—his institute didn't take interns—unless, of course, you were a person who had always felt that the world wasn't enough for them. Dr. Singer, it turned out, admired that kind of audacity. He invited Kamila to his lab in Frankfurt, one of the best in the world. She got her internship and suddenly found herself sitting in her own private space station. It was dark in there, lights blinked, machines beeped. When you touched a neuron with an electrode, it went *sssst*.

Kamila proved useful, she stayed on past her internship, started writing her thesis. She felt she was exactly where she wanted to be—she was already considering her postgraduate options—when her journey took an unexpected turn.

In December of that year, Professor Singer organized one of his famous "Winter Schools," which assembled aspiring and established scientists to Kitzbühel in the Austrian Alps, to talk, argue, brainstorm new ideas, and take them home. Something rare happened that year: a participant dropped out. Perhaps the topic ("Neuro-networks and Synchronization") was a bit much for him.

Huh, Kamila thought to herself, *big science in a winter wonderland. I want in.* And they let her. Winter School turned out to be all she expected, and suddenly a professor named Henry was standing in front of her, tall and slim, his voice soft and engrossing. He spoke about synaptic plasticity. Kamila didn't understand much but listened enraptured.

That night, they stood side by side at the bar. The next morning, they stood side by side on the ski slope. And when they finally said goodbye, they did so with a kiss. They wanted to see more of each other. The plan was to meet in Paris. Until then, while she did her research in Frankfurt and he did his in Rehovot, they would have to content themselves with corresponding by e-mail.

In the days that followed, Kamila, who was researching feline vision in a dark lab, stayed at work until 4:00 a.m. rather than her usual 10:00 p.m. She didn't have wireless internet at home yet, and she passed the night writing down her thoughts and reformulating them. She spent a whole day researching Schrödinger's cat, the famous thought experiment devised by Austrian quantum physicist Erwin Schrödinger, just so she would have something clever to say to Henry. Her coworker warned her not to work herself to death; young scientists should have a life. They didn't know that she was working toward precisely that life. She put more thought and rhetorical force into those letters than all her previous scientific papers combined.

"It's different writing letters to each other," she recalls. "Old-fashioned. Different than today's fast-paced interactions. You really think about what to write, then you think some more. There's something wonderful about all that thoughtfulness." Later, when they married, she bound the letters into a book for him.

But that was still in the future. For now, they lived between Berlin, Frankfurt, Rehovot, and Paris. It was tough at first and eventually became too much. When Henry moved to Lausanne to work for the École Polytechnique Fédérale (EPFL), the famous scientific university where they wanted to build one of the world's finest neuroscience institutes, she followed him, even leaving behind Wolf Singer and her space station, where she had planned to spend a few more years. In Lausanne, she applied to the newly founded Brain Mind Institute of the EPFL, hoping to earn her doctoral thesis.

4

A Strange Boy

Kamila had to sit down.

Where was his mother? Where was his father?

She didn't understand what just happened.

Kamila had rented a boat and rowed out onto the lake. Lausanne bobbed before them. The cathedral, the old city, the harbor, and the people on the banks slowly shrank into the distance. Kai had armbands on, he was only eight, and Kali let her hand glide through the water. Kamila plunged the oars into the water. Left, right—it wasn't easy. She started to work up a sweat. She lifted her head and took it all in. The Alps, the city: still foreign to her but beautiful that day. It was her first excursion with Kai. How would he respond to her?

Everything had gone well with Kali and Linoy, thirteen and ten years old at the time. Henry had spoken to them beforehand. They had listened silently. They went to the movies with Kamila, and Kali cuddled up to her. Linoy kept her distance but flashed her a smile. She was grateful to the girls. She was only twenty-nine. Dating a man with three kids didn't come easily. Kai hadn't come along to the movies. Movie theaters were not his cup of tea. And so, they decided to go for a boat ride. Kamila felt ready. Kali was joining them, thankfully. Kamila planned to make an effort, buy them ice cream, take them both for a swim. Kai loved swimming. What could go wrong?

"Are you sure you don't want me to come?" Henry had asked. "Kai is a bit particular."

"No," she said. "We'll make it work."

She loved children. She was patient and had steady nerves. Kai would feel comfortable.

Sometimes, Henry warned her, Kai threw these tantrums. They appeared out of the blue like a thunderstorm on a hot summer afternoon. Make sure not to overwhelm him.

How would she overwhelm him?

They paddled onward; the sun was shining, the wind sweeping through their hair, the boat coasting. She wasn't so bad at this. The children were laughing, having fun. And then, without warning, something changed in Kai's expression. The look in his eyes hardened, his teeth gritted, and suddenly he was no longer himself. The boy with the armbands took command of the boat. He started shrieking. Kamila didn't understand what was going on. She forced herself to calm down, in the middle of the lake with two kids on board.

"What's wrong, Kai?" she asked. "I'm sure it's not so bad." Kai only grew more frantic, started spitting and swinging his fists. The boat wobbled. The shore was awfully far away, and so was Henry. If only he'd come along! Kamila pretended everything was fine and kept paddling, albeit back toward shore. Kali started quietly talking to Kai, who seemed to calm down but then erupted again, saying things to Kamila that angered her, words she would never repeat, even back on shore, when she recounted the episode. He raged and rampaged, and in her helplessness, she reached out and grabbed him by the nose. She would never hit a child, never attack a child, but it turned out, in panic, on a rocking boat, she could grab this particular one by the nose. Her gesture seemed to say, "Not going to fly, kiddo, not going to fly." Kai was scared stiff. Kali took Kai by the hand, whispered in his ear some more, and Kai seemed soothed by her words while Kamila paddled toward shore, toward his parents, toward firm ground. Kai had a wild look in his eyes, but eventually, after what felt like an eternity, they reached the shore and hopped out of the boat. Now the ground seemed shakier. Kamila had to sit down. Where was his mother? Where was his father?

"My nose. My nose. It's all red," Kai complained when Henry arrived. "Kamila did it."

Henry stared at the ground. He knew what had happened. He would have preferred to spare Kamila the experience.

* * *

"During that first year," Henry says now, "Kamila was just covered in green and blue bruises."

"No, that's not true," she retorts.

"Yes, it is," he says, and she falls silent.

Kai spat, punched, bit and scratched. And she didn't even know why—not the first time he did it, nor the hundredth time. What had she done wrong on the boat? Kai had seemed so excited about the trip. Perhaps he was mad that he didn't get to row or he had suddenly panicked that they were so far from shore. Henry had explained to her that Kai panicked in situations where other kids would jump for joy.

These misunderstandings were commonplace early on. And Kai knew how to get back at Kamila. The little boy had an unparalleled ability to get under her skin. If she walked him to school, he balanced on the curb of a busy street. The looks people gave them! She warned him and eventually grabbed him by the ear. Kai screamed: "I'm telling my mom. I'm telling my dad. My ear is all red."

"Tell your mother. Tell your father," she said, but deep inside it hurt her. She wanted to do better. It was a constant dilemma. You mean well, you try your best, but still you say and do the wrong thing.

Despite all the bruises, she loved Kai. He was an open-minded, warmhearted little person who cuddled and hugged her. He talked to you, you talked to him; mostly you talked past each other, but it didn't matter. He remained lovable. He would tell you all sorts of things, about bowling, about swimming, about the many things he loved. You could play games with him, make music, prepare food together. He wasn't a tender little flower you had to handle delicately; you could take him by the hand and do all kinds of fun things. Bowling, swimming—everything was a hit. So was boating, usually.

Kamila came to understand him better. Henry saw how she brought structure into Kai's life; how she dressed him in the right clothes (the softest sweater); how she patiently made him his cottage cheese sandwich; how she helped him with his chores (not the easiest task for either of them); how she got him in a sleepy mood at the right time, with songs, stories, and the right pillow. Kamila was good for Kai. She became his second mother.

Henry and Kamila spoke about Kai constantly. When they visited a doctor who doubted that Kai was really autistic, she asked

Henry to explain how it all fit together and what else it could be. She started downloading research papers, buying books. She wanted to know everything. This strange boy was no longer a stranger. He was part of her life. It was as Linoy said: Kai was always the center of attention. When he stepped into your life, he ran the show.

Kamila started drawing conclusions. Scientists are like that: They have a question, and they feel compelled to find an answer; that's the fundamental instinct you'll find in every one of them. She soon knew as much about autism as Henry, though she came to it from a different angle. Henry had the outlook of a doctor, an obsessed biophysician. He dwelled on the particulars, delved into the cells, into impulses and molecules. She saw the big picture. She had the outlook of a biopsychologist, a behavioral scientist with a philosophical background. While Henry did his research through the microscope, she analyzed Kai's mimicry, his gestures, his words and fears. Henry knew how electric impulses flowed from one cell to another; she knew how feelings arise in the brain and the effect they have on a person's memory. He saw the lightning strike, the eruption, while she saw the fear, the pain. They were a bit like superheroes who combined their strengths and balanced out each other's weaknesses. We're stronger together, they soon realized. But it was Kai who turned them into a force. He brought something to their research that is often lacking in scientific inquiry: the constant confrontation with reality. Together, the three of them were a force. They were strong enough to go where no one in autism research had gone before: the fusion of life and learning.

* * *

Talking to Henry and Kamila today, fifteen years later, it seems as if they're just starting to grasp the interplay that occurred back then, this act of providence.

HENRY: On the way over here, I was thinking about the influence Kai has had on our life. It isn't that easy because so much just happens—you don't really think about it. Imagine what our life would be like without Kai.

KAMILA: Hmm.

HENRY: Our work has become a shared passion. That has influenced our research.

KAMILA: That's true, but Kai isn't the reason you became a scientist.

HENRY: No, I was born a scientist.

KAMILA: We are both born scientists. We wanted to understand things. It's a compulsion: counting, tallying, and explaining why one thing leads to another. But Kai spurred us on. It couldn't happen quickly enough, even if we were already too late—

HENRY:—and had made every mistake that one possibly can.

KAMILA: We didn't know any better.

HENRY: For sixty years, everyone has said that autistic people have no feelings, no empathy.

(*Kamila is silent.*)

HENRY: When you come at it from another perspective, you reach another conclusion.

(*Kamila remains silent.*)

Henry: The problem is that many scientists have never met an autistic child.

Kamila: They have. They've just never lived with one.

Kai changed everything. Without him, they would have never become autism researchers. Without him, they, the renowned professor and the PhD student, wouldn't have been strong enough. They might have dared to start along another path, but they wouldn't have had the drive to continue when there was no apparent way forward. They knew the established doctrine as well as anyone, the studies, the papers, the assembled knowledge of autism research. They knew the "mind-blindness" theory of autism, the famous experiment that seemed to prove that autistic people can't feel empathy. But Kai didn't behave as an autistic person was supposed to. He contradicted the studies, the doctrine. For a while, they tried to reconcile Kai's behavior with the old consensus; now they tried to reconcile the old theories with Kai's behavior. They dared to do something they never would have if it hadn't been for him: they questioned the old dogma.

5
Doubt

No way was fast enough.
The lab became his home, as it had in South Africa.
Back then, ambition had driven him. Now it was Kai.

Humans feel for one another. Babies smile when their mother smiles. Children often cry when another child cries. That's how we're designed. Children start empathizing with others around the age of two or three. They can surmise another person's thoughts, their intentions, their feelings.

Take this experiment by University of Southern California (USC) psychology professor Henrike Moll. She set up a puppet show for two-year-olds. The Cookie Monster walks on stage jubilantly, places ten cookies in a cookie tin, and wiggles away. A strict doctor comes on, looks into the tin, and says, "There are too many in there!" He takes out eight cookies. When the cookie monster returns, the faces in the audience change: they bite their lips, some open their mouths, a few even try to warn the cookie monster. They are empathizing with him. They know what the cookie monster doesn't know and foresee that he's about to feel sad. Psychologists call this ability "theory of mind."

Empathy is fundamentally human. It allows us to understand each other, to live in a community. It's what makes us social creatures. According to the old consensus, autistic people lack that empathy. They are not social. Take another experiment that involved a puppet show, one staged in 1985 by scientists Simon Baron-Cohen, Alan M. Leslie, and Uta Frith for an audience of four-year-olds who could already express themselves reasonably well. A puppet, Sally, leaves a marble in a basket and exits the stage. Another puppet, Anne, finds the marble, removes it from the basket, and hides it in a cardboard box. When Sally returns, the children are asked where they think she'll look for the marble. "Normal" children say she'll look in the basket. Almost all autistic children think she'll look in the cardboard box. They lack empathy, the experts concluded.

This became an article of faith. The experts couldn't agree on much about autism—so many causes, so many symptoms, hardly any disorder is as multifaceted—but they could all agree that autistic people lacked empathy. This perpetuated an attitude dating back to the discovery of autism, which viewed people affected by it as essentially flawed. The conventional wisdom was set.

This was more significant and tragic than it may sound. The scientific consensus, which saw autism as a deficit requiring correction, has had a profound influence on how the condition has been researched and medicated. Almost every study was built on the same assumption. The Diagnostic and Statistical Manual of Mental Disorders, the bible of psychiatrists and psychologists, which defines and classifies diseases, placed autism in a class of mental disabilities. This guided all further research. Some studies searched for the causes of the deficit. Others tried to remedy it. Hardly anyone considered that perhaps there was no deficit at all.

If anyone did come forward with such a claim, no one listened. If they requested a grant or applied for a scholarship, their application met the same fate as Henry's dossier. Why should foundations support marginal opinions, particularly ones that were so obviously misguided? Even a layman will tell you that autistic people tend to be loners who sit awkwardly in a corner. It is no coincidence that all 625 patented autism medications were geared toward stimulating the brain.

One can assess this in two ways. One might say: autistic people just happen to have that deficit and assuming anything else would be nonsense. That's why research is funded that focuses on it. On the other hand: What if they're wrong? What if all research has been geared in that direction because no one has had the bravery or genius to question a false assumption or fund someone who does? A person like Paul Ehrlich, who at first was blacklisted and then won the Nobel Prize, comes around once a century.

Now, Henry Markram is far from being Ehrlich, but one would be hard-pressed to find a hundred people on the planet who understand autism and the human brain better than he does. And he had Kai at his side. At first glance, Kai seemed to confirm the old consensus, like the times when you called him on the phone and he spoke as if you were in the room, or when he didn't understand your metaphors. On a second glance, however, or on a third and fourth glance, which you get only when you live with an autistic person, you could see that Kai could read you, that he could intuit your feelings correctly. An autistic person shouldn't be able to do that, according to the mind-blindness theory of autism. Naturally, Kai could just be the exception to the rule. After all, four out of twenty autistic children had predicted that Sally would look for the marble in the basket. But Kai may also be something far greater: a well-placed example that calls the rule into question.

Henry wasn't quite there yet. He was still working from the old assumptions when he resumed his research in Israel. Like everyone else, he was searching for a flaw—in his case, the faulty inhibitory neurons that allowed Kai to pet the cobra. If he located the flaw, Henry thought, a medication could be developed, one that offered genuine relief.

In the second phase, he would devote himself to the big questions: Where does autism come from? How can one cure it? There was only one way to do this, in Henry's view: he had to replicate the human brain. It was about time. The new millennium had begun. Just as the rise of chemistry had revolutionized medicine, the new technologies of the age, particularly artificial intelligence, implied a coming paradigm shift. Wasn't it every scientist's duty to take advantage of the latest possibilities?

The Weizmann Institute was too small for this endeavor. Henry needed money, the most capable colleagues, and a supercomputer. It would cost many millions of dollars. The institute promised to provide him everything he needed, but could it really keep that promise?

Henry had an offer from Massachusetts Institute of Technology (MIT). They were offering him a chair and the possibility of grant money. He was on the verge of accepting when an e-mail popped up in his inbox. The EPFL, the university in Lausanne, wanted to become a leader in the natural sciences. The new president was a doctor, a geneticist and neurobiologist, and the vice president was a neuroscientist. They had big plans for Henry.

And so, he drove to Switzerland; it was the year 2000, shortly before he met Kamila. He introduced himself to his prospective new bosses like this: "I have an autistic son. And I think neuroscience is too divorced from real life. In my opinion, the best way to really alleviate this suffering would be to simulate the human brain. That would be my plan for this institute. I should also mention that I would not be instantly available. I first need to continue my autism research. I would need another two years to lay the foundation for such an institute. I would have to find the right people and collect fundamental data."

"Okay," they said. "We'll wait."

In 2002, having finished up in Israel and recruited a team of collaborators, Henry took them up on the offer. "I am ready now," he said. "I could go to MIT, but I'd prefer to build something from the ground up. I will, however, need a supercomputer. It costs twenty million dollars." The Swiss listened, stone-faced.

"Yes, that works," they said.

The work began, or rather the preparation. It would take three years before he could even start what became known as his Blue Brain Project and another eight years before it would evolve into the Human Brain Project. Even bigger, even more unbelievable: the EU pledged a billion euros of research money. This made Henry the most prominent neuroscientist in the world, admired, envied, hated. It didn't matter much to him. He was far removed from the motivations of academic life and was no longer concerned with acquiring knowledge and prizes. It was all about Kai, who sank more deeply into his own world by the day. Kai wouldn't give Henry a round of applause when he got home. He was more likely to insult him because Henry had bought the wrong Christmas present, ordered the wrong food, or used the wrong words. This weighed on Henry, who day by day—on top of managing the brain project and doing his work as a neuroscientist—sunk more deeply into his autism research. Kamila was concerned to see him sleeping only four hours a night. His colleagues grew nervous when he hustled through the hallways, a restless spirit leaning in to inspect their data. It couldn't go fast enough for him. C'mon, let's give it another try—again and again. The lab became his home, as it had in South Africa.

Luckily, they still had Anat, who lived nearby and got along well with Kamila. Thanks to her, Henry and Kamila were free to spend all day—and often all night—steeped in research. Kai lived with her. She was always there for him, whether he was hungry, sick, or throwing a tantrum. She gave him unyielding love and comfort. She was there when others were not. She was, as Kai said himself, the best mother in the world. Henry and Kamila were his confidants on the weekend. They were responsible for bowling, hiking, the—structured—escape from everyday life. "What should we do?" they asked and offered Kai twelve suggestions. In the end, he liked them all and they just went bowling.

6
Don't Cross That Line

"Don't get your hopes up," the principal said.
"An autistic child will never be normal."
What do you know? *Henry thought to himself.*

"Please take a seat."

Henry sat down.

"I'm glad you've chosen to enroll your child with us. I'm sure you have questions."

"You have a good reputation," Henry said. "You have experience with autistic children. Now you've met Kai. What's your impression of him?"

"It's too soon to judge, but I think he'll be in good hands here. As a private school, we can offer Kai a lot that other schools can't afford. We adapt our approach to the child's needs. Each student has their own teacher."

Henry and Anat were pleased and nodded. After so many false starts, they had spent a long time debating what school would be best for Kai.

"Also, we teach according to a special methodology. The ABA method. The child learns through rules, rewards, but also through sanctioning."

"I see," said Henry. "We didn't exactly have the best experience with that at the Montessori school."

"Montessori?" The teacher raised an eyebrow.

"Yes, and Lynda Thompson from the neurofeedback clinic."

"Neurofeedback?" She raised her second eyebrow.

"Yes. Lynda said there were three ground rules: be nice to him, be nice to him, be nice to him."

"Well," said the teacher, "the ABA method was developed specifically for autistic children. And studies have shown that it works: whether it's intelligence or language acquisition, the children develop more successfully than in generic therapies."

Henry and Anat stared at her vacantly. They had their doubts—and so did Kamila, whom they had consulted at length. Then again, they weren't teachers. They had to trust someone. They faced the same dilemma as every parent with an autistic child. There are so many forms of autism, so many approaches to therapy. How do you proceed? Whom can you believe?

* * *

Lynda would have wrung her hands if she'd known that Kai was going to be taught according to the ABA method. "Behavioral therapy like ABA doesn't work for Asperger's," she says. "It's geared toward symptoms. If a child avoids eye contact, they force him to do exactly that. That isn't good for the child. If you want a child with Asperger's to open up, you should try to approach him through his preferences. I call that participation and redirection. For example, if you want a child to learn to read, you should find out what interests him, what excites him, and give him books on that topic."

The successes of ABA that the teacher mentioned had been achieved only with classically autistic people. And those successes were controversial. Certainly, some parents had good experiences. Others felt their child was being trained or conditioned. The child was doing what the teachers wanted only because he or she had been drilled to do so. This didn't seem to make them happy. How well would it work for Kai? Henry felt at the mercy of chance, like all the other parents. He had yet to reach his own conclusions—suggesting that there is a therapy that works for all forms of autism.

They decided to put their trust in the teacher. If only they had trusted their inner voice instead, as Anat usually did. What the teacher said on their way out should have been enough of a red flag. "Don't get your hopes up," she said. "An autistic child will always be autistic and never be normal. Kai will never be able to do certain things." She didn't say this in bad faith. She meant well; her tone was considerate. Henry and Anat could hardly believe what they were hearing.

What does she know anyway? Henry thought. He was a scientist. He believed in progress. He would find a way. Kai would go to work someday, find a wife, and live a normal life, like everyone else. There is no false hope. Hope is good.

* * *

Their doubts solidified, but did they have a choice? Kai had to go to this school.

They had tried a different one before, an Israeli school. They'd assumed the teachers and students would speak Hebrew and hoped this would make it easier for Kai to adapt, but unfortunately, they were mistaken. This was Lausanne, and everyone spoke French, of course. Kai sat there and didn't understand anything, and the kids bullied him, pulling his hair, pouring water down his pants, laughing at him. Kai started hiding in the restroom or jumping out of his seat in the middle of class and walking around, yanking his classmates' textbooks away. If he couldn't learn, they wouldn't learn. He wanted the other kids to play with him, to pay attention to him. He wanted to be the center of attention. The principal kept calling Henry and Anat. Their pulse rose every time the phone rang. Eventually, he said, "Kai is a nice boy, but we can't take responsibility for him anymore. The other children have rights too. We can't keep him in class."

Autistic children often get expelled from school:

Because they are wired differently and tend to disturb the smooth hum of operations.

Because they need small classes, as well as time and space to be alone.

Because they might need to go for a walk during the second period to get away from the noise.

Because they can't sit still, and they hold their ears shut.

Because they become stubborn when something changes suddenly, like a class, or a classroom.

Because they scream and bite and spit.

Because teachers aren't prepared for children like that.

After Kai's expulsion, Henry and Anat interviewed at a lot of other schools. "I can't expect that from my teachers," one principal said. Others rejected Kai with no explanation. Only a school for the mentally challenged was willing to accept him.

They had no choice but to ignore what the teacher had said and pay a steep tuition for a questionable teaching method.

Henry had heard stories about ABA schools. He heard that the children had to eat things they didn't like, that they had to remain seated when everything compelled them to move, that their rituals—the very things they loved most—were driven out of them. If they stopped playing with the water, they got to eat dessert. If they stopped placing cards in a row, they got their teddy bear back. If they wanted to always keep their teddy by their side, it was taken away from them. If there was a special person they always wanted to keep close, they were allowed to see them less. Instead, they had to solve tasks, recognize faces, stand still, and talk to strangers, and they were praised excessively and rewarded if they succeeded. They were trained like dogs, opponents of the method said. Henry repressed these thoughts.

He had no way of knowing that this form of education contradicted everything he would come to discover in his own research. ABA was as wrong for Kai as it could possibly be. It was the next mistake in a tragic string of them, mistakes made only because they wanted to be the best possible parents. Kamila says:

> The school itself was quite good. Everyone had their own teacher. But the ABA method was not good. These children are comforted by their rituals. The rituals represent something they do well, something they understand. When you take that away from them, they panic. According to our theory, you should let your child

indulge in its rituals, play along with them. Accept their world; delve into it. If you do, they will come to you.

This is especially important if your child has particularly severe rituals. Kai isn't so extreme; he speaks, interacts. He's not one of those kids who stop talking altogether, who just sits alone and builds LEGO towers. It's especially bad for those poor kids if you try to take away their rituals. To give a good example: Ron Suskind wrote a bestseller about his experiences with his autistic son, which was made into an Oscar-nominated documentary. He visited me and Henry while making the film, interviewed us about our research. Suskind's son watched Disney movies from morning till night. That was his world. He was reclusive and didn't talk. The doctors advised Ron and his wife to forbid their son from watching those movies: he had to do something else. The Suskinds tried to take him away from his movie collection and into other contexts. They did it with the best intentions, because that's what authorities in the field said was the right thing to do. It turned out not to be good for their son. He felt comfortable watching those movies and learned a lot from them.

Eventually, Ron learned something remarkable: if he spoke like a Disney character, his son could talk to him. The Suskinds managed to connect with their son through this Disney world. They could start a conversation with him because they had accepted his comfort zone as the foundation. That was the revelation: "Okay, I first have to enter your world, and then, slowly, I can show you another world."

* * *

Kai stayed in that school for a year. The kids had all forms of autism. Sitting them all down in a class together was impossible. One kid could crunch numbers like Einstein, while another couldn't find the classroom by himself. One sat quietly in the corner, while another talked nonstop. It was good that Kai had his own teacher. That is not to say that the man did everything right. When Kai had to go to the bathroom, the teacher wanted to guide him there by the hand. Kai hated that and showed it. The tasks became stricter, the punishments harsher, the rewards more seldom, and Kai grew decidedly more strange. He picked up habits from other children. One of them threw himself to the ground; suddenly Kai started doing that too. Henry thought this would pass, but it didn't.

They drove him to school in the morning with growing reluctance. In the afternoon, they picked him up again. When Kamila was on pick-up duty, she had to take the subway because she didn't have a driver's license. Kai got a kick out of testing her boundaries. He no longer just teetered on the curb; he marched right into the street.

"Kai! You can't walk into the street."

He just laughed.

Kai knew just as well that he wasn't supposed to cross the white line on the subway platform. He looked up at Kamila—he was cute, those two big eyes, that mischievous smile—as if to say, "I'm going to do it."

One foot over the white line.

"Kai! That line is off-limits!"

His second foot crossed it.

"Kai! Don't cross that line!"

He stepped to the edge and teetered. People looked at them, went "Oh!" and Kamila grabbed him by the ear. Kai screamed!

"You listen to me!"

He screamed louder than you can imagine. The people gawked, though Kamila had grabbed him gently. They boarded the train, where he kept crying.

"I'm telling my mom."

"Good, tell your mom."

When he realized this wasn't working, he started swinging his feet back and forth, looking at her and swinging, until he hit her kneecap.

"Kai! Stop that!"

He kept swinging his feet and bumping into her, pestering her until they got home. The little schemer knew exactly how to push her buttons. Kamila was on the verge of flying off the handle, but she didn't let herself. *All kids are like this*, she thought. They throw themselves to the ground in the supermarket. In their terrible twos and thrashing threes, they say "No!" to almost everything. The difference was that Kai was still doing this at age eight. Indeed, he was doing it more than ever.

She tried her best to recall the golden questions: "What does Kai like? How can I understand his world?" Music, bowling, pizza: that's how she made friends with him again. She wouldn't let him go for anything in the world, that cheeky devil who melted her heart when he cuddled her, helped her cook, helped her shop, and told her the funniest things. He saw the world so differently.

7

Back in the Land of the Special People

Kai smelled the air, felt the sun on his nose.
It was like he had never left.

The day came when Anat decided it would be best for Kai and her to return to Israel. Kai felt foreign in Switzerland and so did she, her marriage to Henry was ancient history, and all her friends and relatives lived back home. Kai was the only thing keeping her in Switzerland. And while this country was wonderfully beautiful—the air pure, the streets clean, the people so polite—it wasn't welcoming to him. Very few countries in the world would welcome someone like him, at least not to the extent he required. Autistic people are foreign almost everywhere.

* * *

In 2006, the UN adopted a treaty that would become law two years later. It enshrined the rights of people with disabilities. This was necessary because people everywhere like to forget that disabled people are human beings. They are excluded, tormented, or pitied. The convention seeks to protect their human rights and fundamental liberties. More than 175 countries have signed on so far. One of its most important tenets: People with disabilities have the right to participation, the right to be included. They have the right to live like everyone else. They are part of society and should not be hidden away in asylums, special schools, sheltered workshops. They shouldn't be excluded at all. They should be where they belong: among us.

The United States has not yet ratified this convention. Switzerland was one of the last countries to do so. And not without a public debate—mostly about what this inclusiveness would mean for Swiss schools. Germany implemented it earlier but hasn't made much progress. UN inspectors traveled to Germany to keep tabs on its implementation. In diplomatic language, they concluded that they were concerned:

- "that the measures to dismantle stigmatization, particularly of people with psychosocial and mental disabilities, have been without effect.
- "that parents of children with disabilities cannot freely decide what education their child receives.
- "that a majority of children with disabilities attend special needs schools." Seventy percent was the actual number.

This describes the status quo in most progressive, rich countries. The fate of autistic people in the developing world meanwhile is unfathomable.

There are, however, a few countries that treat disabled and marginalized people in a smarter and more humane manner. One of those countries is Israel. Eight out of ten Israeli children with disabilities attend a "normal" school.

HENRY: Israeli society is simply more accepting of diversity.

ANAT, *laughs*: So many of us have a screw loose, it hardly stands out.

KAMILA: Children are welcome everywhere. Even if they have a tic of some kind, it's totally fine. In Switzerland, it can be nerve-racking bringing a child to a fancy restaurant. If they have a tic, you could just be kicked out. Other guests and waiters look at you funny, and you start to think, *My god, I'm such a bad mother.* It's different in Israel. We all agreed it would be better for Kai there. Particularly when he's older and wants to get an education or a job. Israel has invented jobs for people like him.

Israel is a nation of immigrants. In its schools, you'll find children from around the world, including many refugees who arrive underprivileged and become part of society. Inclusion is a signpost of an evolved culture. Israel does many things differently for people who are different. Tourists in wheelchairs are often surprised at how easy it is for them to get around. Companies with more than twenty-five employees are obligated to hire people with disabilities. In the digital economy, a whole sector of companies has popped up that develops apps for blind, deaf, or autistic people. When people with disabilities protest for higher pay—as they did in 2017—people show solidarity, newspapers report on it, and even the police offer food and drink. The whole society cares about "people with special needs" and does not consign them to the status of welfare recipients. They allow these citizens to give something back to society. Even the army employs people with autism. In a division called Unit 9900—known popularly as the Eye of

Israel—they have an important task. They evaluate maps and satellite imagery. Thanks to their photographic memories, these people locate troops, weapon depositories, and rocket launch pads that others can't see. They're admired for this, their achievements acknowledged. This carries over into the athletic world too. It's no coincidence that Israel regularly achieves better results in the Paralympic Games than in the Olympics. Since 1960, they have won 380 medals in the former and nine in the latter.

Why, then, should Kai torment himself in Switzerland, where, as Kamila said, an asylum rather than a school awaited him? The Markrams made a painful decision: Kai would return to Israel with his mother. Kali, who was so important to him, who also felt constrained in Switzerland, would join them.

Linoy, thirteen at the time, was old enough to decide for herself. "It was the hardest decision of her life," says Henry. "I admire how she managed it." She thought about it for a long time and ended up staying.

Kai cried upon leaving, everyone cried, they held each other close. They pledged to see each other as often as they could. During the holidays, Kai would fly to Lausanne, and during the school year, Henry would come to him. Soon every one of them had a frequent-flier card, including Linoy and Kali. They wanted to see each other, wanted to see Kai, wanted to see Mom, Dad. Many threads and strings kept them connected, with Kai at the center.

Their patchwork family continued to grow. Kamila and Henry had two daughters, Olivia and Charlotte. Sometimes when the parents traveled to conventions, Anat would travel to Switzerland to look after the house and children. It's been that way for years. Kai, in the meantime, became a teenager and young adult. Much has changed, but one thing hasn't: no matter how often they get together, Kai and Henry miss each other. "We call each other then," they both say.

* * *

Kai smelled the air, felt the sun on his nose. It was as if he'd never left Israel. He rambled through the neighborhood, striking up a conversation with everyone. He made friends in his own strange way. He seemed so free and happy. Unfortunately, he had retained his fear of school. Where would his parents send him?

As is standard in Israel, a regular school offered to take him. A friend of Anat, a teacher, advised against it. "That's not a good idea," he said. Even if people in Israel are open-minded and the teachers are well-trained, it's no paradise. Parents have to weigh the pros and cons. Kai was so naive, he believed everything and showed his feelings openly. Anat remembered all too well how his "friend" had incited him to throw stones at cars. "I'm afraid the other pupils will take advantage of him," Anat's friend said, "that they'll make fun of him."

Anat and Henry took stock. They were thinking less about Kai's educational prospects and career opportunities. They just wanted him to feel good, to recover some of the self-confidence he'd lost in the United States and Switzerland. They wanted him to be surrounded by good people, for there to be no rift between him and others. This was so important to Kai, who more than anything else wanted to feel included; who, long before the UN had added "participation" to their definition of equality, had only one desire: to be among people.

With that in mind, they sought out and eventually found a school for children with learning difficulties—a special needs school. The classes were small, the teachers loving, the children happy. It seemed counterintuitive. They had moved Kai to Israel because they wanted him to live in a more inclusive society, because autistic people were treated like humans there rather than outsiders. And still they ended up sending him to a special needs school. Now they were the ones secluding him. But it wasn't really counterintuitive: it was as logical as it could possibly be. They listened to their hearts, their inner voices. They were free to decide. This shows how forward-thinking Israel really is on these matters. While the UN has complained that parents in Germany aren't free to decide what schools their children attend, parents in Israel enjoy that freedom to the fullest. Yes, you can include them; but you don't have to. And special needs schools aren't bad, per se; a study by the University of Lucerne praises the advantages that this form of schooling can have. For many, it's a blessing. Blessed be a society that offers both: normal schools as the rule, even for pupils with disabilities, and special needs schools as a necessary exception.

Kai felt comfortable in the new school straightaway. For the first time since kindergarten, his parents could breathe freely. No drama, no calls from principals. Sure, he might have learned more elsewhere, the material was easy, and Kai was downright lazy, but that was beside the point. He was good at drawing, good with computers, good in music class, and excelled at basketball in PE.

After school, he hung around in the schoolyard, simply because he felt like it. No one had to pick him up or pull him back onto the sidewalk. He didn't throw himself on the floor anymore.

Occasionally, he did complain: "No one likes me. They're all weird." When Anat called the school, concerned, the teacher said: "What? That's nonsense. He's the king here. He's never alone during break." Kai couldn't always see all the love people felt for him.

He bloomed in Israel. All the houses in the neighborhood were open to him. You could walk into your neighbor's house as thoughtlessly as you did into your own, without advance notice. No one was too young or too old; everyone came together. Kai was part of a community. He was included. Only his search for best friends remained clumsy. In fact, his mother found friends for him, by inviting over parents with kids his age.

* * *

When Kai hit puberty, the old worries resurfaced. Anat couldn't manage his social life anymore. At that age, your mom can't invite friends over for you. The contrast between Kai and his peers grew starker. It was just like in first grade, when all his friends had matured into schoolchildren and he remained a kindergarten child. This time, his friends became teenagers, giving in to all the adventure and teen mischief—cigarettes, beer, love letters, moped joyrides, hip-hop, and mohawks—while Kai still wanted to read comics with them or play kids' computer games. Soon they were strangers. Sure, they still said hi, but no one was about to invite him out to a club.

In school, the teasing increased. He stopped going to basketball, no longer hung around in the yard after school. In class, he flipped a table. The old fear gripped Henry and Anat. But Kai pulled himself together. His teachers reported that he was no longer as prone to tantrums, that he even managed to ignore the other kids' taunts. He was most friendly to girls. He wore baseball caps and jeans with holes in them. It hurt him deeply when no one wanted to go on a date with him. But he didn't complain. At home, he reported that he had the most girlfriends in school. Of course, he did.

8

Tania Can't See It

Henry was even paler, his pronouncements more clipped.
Were they wasting their time?

Tania first met Kai when he was ten years old. Henry and Kamila had invited their colleagues from the university over for dinner. They ate duck and looked out over Lake Geneva, and the conversation sparkled like the champagne in their glasses. Everyone had brought their partner. Tania was with her boyfriend. A PhD student, she was the youngest in the group. She had been working for Henry for two years. He had never mentioned Kai.

The ten-year-old was slender and lovable, she recalls, and anything but quiet. He knew how to draw attention to himself. He wandered around and started a conversation with everyone, about bowling and swimming, about himself and his family. It was a pleasure to listen to him. What a bright boy, she said to herself. On the way home, she was surprised to hear her boyfriend say:

"Now, that's a weird kid."

"Weird?" Tania replied. "He's just a child."

"No, no," her boyfriend said. "Something isn't right there."

Tania didn't give it any more thought. On Monday, back in the lab, she thanked Henry for the lovely evening; she had already forgotten her boyfriend's comment. Henry didn't follow up either. He was never one to chat about his son's autism.

In retrospect, Tania is glad she didn't know. The doctoral student has since become a professor who now works in Basel. "If I had known, the pressure would have been too great," she says. Perhaps she would have rejected the job, turning down the man who showers her with praise until this day. "Henry isn't everyone's buddy, but he impresses me a lot. Every time I run into him, even for only five minutes, I'm once again astounded how smart he is." Henry should consider himself lucky that Tania took the position, for she would give his research a decisive push.

* * *

Tania had no idea what awaited her when she stepped into the dean's office. She was one of the smartest students in the country, had won prizes in chemistry and quantum physics, but her dissertation topic wasn't satisfying to her. She told the dean she wanted to transfer. Didn't the EPFL have a new science department? Might there be something for her there? "Well, yes," the dean said, nodding his head, he might have something for her. The university was expecting a remarkable professor. Neuroscientist Henry Markram—she had surely heard of him—had been courted by the world's best universities but ended up choosing their institute. Markram had big plans and could certainly use a gifted person like her. Could she imagine transferring to neuroscience? He knew she was a chemist, but wasn't the brain mostly chemistry, anyway? Her prize in physics had proven her versatility.

The dean gave her Henry's telephone number, said she could say he referred her, and soon she was sitting across from the new star professor. Five minutes into their conversation, she thought: *This is what I want to do.*

Henry didn't even look at her résumé. "When can you start?"

Two months later, she was sitting in one of the best labs in the world, surrounded by vials and test tubes, centrifuges and microscopes, the best computers and tools. Henry told her that he wanted to go further than any autism researcher before him. This meant not only measuring brainwaves and counting neurons but delving as deeply as possible into the cells and understanding how they communicate. Through this precision work, he wanted to get to the bottom of Merzenich's thesis and closely study the faulty inhibitory neurons.

She would be responsible for creating an animal model—breeding autistic rats. He had already done the groundwork. Mercury, alcohol, and the drug thalidomide didn't seem to do the trick. But there was this epilepsy medication. She could take a look at it and feel free to make recommendations of her own. Tania took off to the library and pored through the relevant books. She went to the database, scrolled through the right pages, and eventually came upon a frightening study. Some people who suffered from epilepsy had been prescribed a medication called Depakote, which contained an agent called valproic acid (VPA) that seemed to do a good job alleviating seizures. In pregnant women, however, it had proven to have horrific side effects, one of them being that nine in every hundred newborns developed autism.

Many years later, in 2017, France would establish a fund to compensate women who had been prescribed the drug during pregnancy.

Scientists had since discovered that VPA also caused autism in rats, which closely resembled the human disorder, causing reclusiveness and phobias. To induce the disorder, one needed to administer the medication to the animal on the twelfth day of pregnancy, when the neural tube closes. Tania rushed into Henry's office: "I've got it!"

The work began. Tania had her difficulties with the patch clamp, the method of measuring a cell's electrical currents that had earned Henry's mentor Bert Sakmann and his colleague Erwin Neher the Nobel Prize. It required great skill. She had to work with a tiny artificial hand, a micromanipulator, which translates crude human movements into fine mechanical ones.

She took one of the rat brains, which lay there like a loaf of bread, and started cutting razor-thin slices off it, 0.3 millimeters that contained a million neurons. She submerged the slice in oxygenated brain fluid to prevent it from dying and then used a glass micropipette to stab into the brain cell—into the inhibitory neuron. The procedure calls to mind images of artificial insemination, when the sperm is inserted into the egg, but the brain cell is a hundred times smaller. A tiny flake hung on the end of the pipette, a membrane patch. With tiny hand movements and big physics, you could use it to measure the cell's electric currents, monitor how it fires when exposed to stimuli. In the lab, electrical impulses provided this stimulation, rather than, say, a dancing cobra. If the cells proved to be talking to each other less, this would bolster Henry's thesis. Perhaps a medication could be developed to remedy the flaw. It was important research: Tania was inspired by the task. Now *this* was a dissertation topic!

Working on a cell is an achievement all by itself. Scientists spend whole workdays "hunting them," trying to "catch" a single one at the right angle. Henry, on the other hand, had developed a technique and a certain dexterity that allowed him to measure up to twelve cells at once. This allowed him to monitor not just the activity of one cell but the communication between several. He could see whether one cell was talking to other cells and if the other ones could hear it.

Henry taught Tania the technique and their work progressed. She cut slices; adjusted the microscope, an inverted one that looks up from below; spent hours adjusting the micromanipulator; and stabbed into the cell, measured its currents, and compared the numbers to old studies. It seemed almost jinxed: the values suggested nothing unusual. Measurement after measurement, cell after cell, everything appeared normal; the inhibitory neurons were doing their job. Meanwhile, in the cerebral cortex, where the higher cognitive functions are located, the cells responded more intensely than usual. This was the exact opposite of what they had predicted. Weird: an autistic brain that overreacts. What on earth was going on?

Henry kept looking over her shoulder. Were the measurements correct? Had they made a mistake? Was the gauging station programmed correctly? Had she used a vibration isolator? Was the Faraday cage correctly arranged, blocking off outside currents and friction that could skew the results? Was Tania having trouble with the subtle handiwork of the patch clamp?

Tania's anxiety swelled as the months passed. The dean awaited her dissertation, while Henry's looks grew more concerned. She had nothing to show either of them. High-flying Tania was stuck. "Look for another topic," her lab advisor, Gilli, suggested. "Nothing will come of this."

Henry, her mentor, showed his face less and less. When he did, he was even paler, his pronouncements more clipped. Why should he waste more time on these rats! Did they need another animal model? Another thesis? Another PhD student? What could he change? Things couldn't go on this way. Henry had lost four years to these inhibitory cells. He had blown his first research grant on them in Israel, frittered away two years of research in Lausanne, and he had precisely nothing to show for it. Four damn years had passed since Kai had petted that damn cobra, since Henry had sworn to dedicate his knowledge to something useful. In those four years, he had made no progress whatsoever. In the meantime, Kai had regressed considerably. Henry found it harder than ever to reach him in his world.

9
Last Chance

After months of failure, Tania stood there,
her heart beating in her throat.
Was this the breakthrough?

Where were you when the terrorists flew into the Twin Towers? At home? On vacation? In the office? Who were you with? Almost everyone remembers those details. On the other hand, hardly anyone remembers where they were on September 7 or October 23. That has to do with the way our brains work, with the amygdala—really there are two of them, two amygdalas—two small areas in the cerebral cortex that steer our feelings. The amygdala is where our fears come from. It's where dangers are evaluated. When you're in shock, experiencing fear or grief, your amygdala starts overperforming, sending alarm signals to your memory. You never forget that feeling of alarm. It protects you from getting into that situation again, reminds you to avoid that danger in the future.

The amygdala was Kamila's field of expertise. The molecular biologist/behaviorist was writing her doctoral thesis about it. Her work had nothing to do with autism, but now Kai was there, and again, he didn't step into your life without changing it. Soon, Kamila was reading more about autism than any other topic and following Tania's work as if it were her own.

She and Henry discussed it frequently, and they were on the brink of aborting the test series. "Let's give up," Henry said on more than one occasion, but then he wavered, sensing that the solution was right under their noses, staring them impudently in the face. "It was almost like it was hidden behind a Venetian mirror," Henry recalls.

* * *

And then the day came when he really wanted to give up. "Let's leave it at that and try to find another way," he said to Tania.

"No," she responded. Her dissertation would be doomed. "Let me try something else," she begged. Henry agreed.

Tania proceeded to do something crazy, and she did it for no reason she could explain, a wild stab in the dark. After spending two years probing inhibitory cells in vain, she decided to try their opposite number, the excitatory cells, the ones that remind you to move your hand off the hot stove, the ones that agitate the brain rather than calm it down. She stood there, alone in her lab. She stabbed the micropipette into the patch, stimulated it, and could hardly believe what she was seeing through her microscope. The excitatory cells perceived the stimuli as twice as strong—they talked more to each other; they were really blabbering. And, to keep it in colloquial terms, they had a lot more followers too, many more cells receiving their messages. The excitatory cell of a normal rat connects to ten other cells. The excitatory cell of an autistic rat has twenty such connections: a firework of signals, doubly as fast, visible from twice as far away. Miracle cells.

After so many months of perceived failure, Tania stood there, her heart in her throat. She went straight home. She didn't mention anything to Henry and Kamila. She might have made a mistake. But the next day she achieved the same results and the following day too. Finally, on the third day, she approached Henry.

Everything is amplified? Henry was baffled. They redid the experiment, patching and clamping like crazy, and got the same result: it wasn't a fluke. How on earth could that be? Everyone spoke about a deficit, but they had found an excess, a strength not a weakness. These high-performance cells weren't connected to each other by the usual streets but by veritable signal highways. Impressions and perceptions sped through them. Everything the rats saw, heard, or smelled was amplified: their brain amplified it. They didn't have a deficit of feelings; they had a surplus.

Henry, the traditional scientist, would probably have aborted the series of experiments right there. He would have reviewed the past research, the insights of the past decades, and concluded that, for whatever reason, they were on the wrong path. He would have started again from square one. If it hadn't been for Kai, that is. The results clashed with the consensus but fit Kai perfectly.

Henry and Kamila couldn't contain their excitement. If, contrary to expectations, the cerebral cortex was disproportionally active, the impressions were amplified, and the autistic world was faster, louder, more colorful, how did it look in Kamila's area of expertise, the amygdala, where memory and feelings are located? Would those areas be amplified too? "Imagine that," Kamila said. "Then everything in the books would be wrong."

Kamila dropped in on her thesis advisor, Carmen Sandi: "Could we perhaps expand my postdoc? I know of a project, one related to autism. Tania, over in the lab, has discovered something that calls everything we have been taught on the subject into question." Once she had told Carmen more about Kai and Tania's breakthrough, her advisor said, "Now, doesn't that sound exciting? Yes, gladly." Kamila recruited Tania into her team. Together they would take a close scientific look at the rats' amygdalas, monitoring their fears, feelings, and memory. It was a complete break with tradition.

Autism research has always relied on experiments with animals, often monkeys. These experiments often involved removing the amygdala from a monkey's brain, whereupon it would lose its drive and position in society and be left cowering in a corner, no longer interacting. The monkey has autism, the scientists concluded—they interact less. The trials began, inevitably leading to treatments that were geared toward stimulating the brain, awakening feelings, animating them, bringing them back to life.

Kamila and Tania did the opposite. Their thesis: If these impressions were racing through the cerebral cortex, it had to have an impact on the rats' amygdala, creating an excess of emotions, feelings, and memories. Tania would study the cells; Kamila, the biopsychologist, would monitor how the feelings moved through the brain, how their pain and fears traveled between the amygdala and the cerebral cortex, and how it affected the animal's behavior.

They first devoted themselves to the cells. They patched them, stimulated them, and watched as the cells responded. In the amygdala, where everything was supposed to be dead or weakened, the needle went up. They were on the right path.

They now inspected the animals: What are they learning? How do their memories work? How do they process fear? They appropriated a famous experiment known as the water maze.

Kamila explains: "You sit a rat down in a pool of water. Rats can swim. There is a small island in the basin where the rat can rest, but it doesn't know where it is at first. The rat swims, it's under stress—it takes time to find the island. You record how long the rat takes to find it. You take the rat out, put it in a second time, and track how long it takes to find the pedestal again."

Kamila placed "normal" rats and autistic rats in the basin. The result was surprising: the autistic rats found the pedestal more quickly. They learned faster and had a better memory. In the next iteration, she injected fear into the experiment, the rats' personal 9/11, so to speak. She wanted to find out how fear got lodged in the animal's memory and how it affected their behavior. The rat sat in a box, on a grate, and a shrill sound rang. After twenty seconds, Kamila gave the animal a small electric shock through the grate, less than a millivolt but still unpleasant. She repeated the test, and the rats learned that the alarm predicted the electric shock. They ran around, sniffing, until the sound rang, at which point they froze.

Then Kamila changed the setup of the experiment. It seemed the same: the animal ran around, sniffing, the sound rang, and it froze in fear. Only, this time, the shock didn't come. If the rats froze quickly and remained that way for a long time, it indicated they had good memories. Again, the autistic rats defied expectation: not only did they learn more quickly, they also had greater fear and were better at remembering it.

After further attempts when the alarm sounded but no shock followed, the normal rats learned that there was no danger anymore. They didn't get worked up. The autistic rats were scared for longer. They often sat frozen for the entire eight minutes of the experiment. Their fear was greater and didn't go away!

Wow. Kamila could hardly believe it. Simply everything was amplified: The eruption in the cells that Tania was monitoring. Its effect on the rats, their learning, their fear, their memory. This contradicted the mind-blindness theory of autism, contradicted the research they had found in old scientific articles and books. It corresponded exactly, however, with what they witnessed in Kai, who seemed to recall every minor trouble he'd experienced, who could remember years later what chair he was sitting on when they forced him to eat that disgusting piece of lettuce. He couldn't erase a minor hiccup from his memory.

Suddenly, everything made sense. Kai had no deficit whatsoever. He didn't feel too little. He felt too much. His withdrawal was not the disorder—it was a reaction to it, to his personal 9/11 reoccurring on a daily basis. If you want to help him, you have to act on his strengths, not his weaknesses. Suddenly, a common footnote they had seen in many studies appeared in a new light. The sensitivity of autistic people was well-known. However, it was consistently treated as a marginal aspect, a moving parenthetical, like the high-functioning savantism popularized by the film *Rain Man*. But this sensitivity was not trivial; it was the key.

10
The Tree

This is how Kai must feel. Everything is amplified.
He lives in an unbelievably intense world.

Henry and Kamila were itching to share what they had discovered, but the scientific community didn't want to hear it. Kamila's dissertation, containing all their research of the past few years, earned a silent reception. Who reads dissertations, anyway? She rewrote it in essay form and sent it to *Science* and *Nature*, those venerable science journals that are read by so many. *Science* rejected it out of hand; they mistrusted a paper that contradicted the long-established consensus on autism. *Nature* wavered and ended up passing it on to a sister publication that specialized in neuroscience. That was a compliment; they saw value in her work. But who would read this journal?

When Kamila was traveling and wanted to look up something in her own paper and download it, she saw that the magazine's subscription cost thousands of euros. Not even EPFL, one of the fifteen biggest universities in the world, was willing to spend that. The most exciting autism study of the past decades was buried in a remote archive.

Well, that wouldn't do. They would write a new paper, one for laymen not experts. And it would have to be freely accessible. To that end, they founded a publishing house of their own, a platform and a website, where scientists from all over the world could share their knowledge. They would launch this venture by publishing a magazine, which would include an essay about their breakthrough. For months, they hunched over their documents, summarized them, editing, cutting. The main takeaways of the complicated scientific text should be understandable to everyone: parents who were as helpless as they themselves had been; prospective students who were considering neuroscience; and everyone who was open-minded and didn't restrict themselves to conventional knowledge.

"Let's take a trip somewhere," Henry suggested one day. He wanted to get away from the fluorescent lights and office air, to a place where they could see clearly and breathe freely. You need to move to write with verve. "Why don't we go to South Africa?" he said. He was referring to the Kalahari Desert, his roots, the places of his childhood. He had always wanted to show them to Kamila.

Kamila was excited. Henry had told her so much about the Kalahari, about their farm, the vast steppes, the landscape's rich color scheme. As their trip drew closer, his tales grew taller—the mountains higher, the houses more splendid, the savanna more colorful.

In the seventeenth century, Henry's family owned wide swaths of the Kalahari. Their property shrank over the centuries and generations, though during Henry's childhood it was still so vast that it took a whole day to drive from one end to the other. His family sold it all when his grandfather died. The new owner had turned it into a nature preserve, with water buffalos, black rhinos, and black-maned Kalahari lions. They would visit his childhood home, which, judging by Henry's stories, must have been something of a palace. They would see if the old barn was still standing, where Henry had always hidden the Karakul lambs, those soft, black curly-haired sheep that his grandfather bred and slaughtered. And they would drive into the savanna to see the thorn bushes and water fields.

In Johannesburg, far from Henry's home, they took off in a rental car, feeling like their minds had yet to fully arrive in the huge country. Preoccupied by their paper, they managed only to glance at the passing animals and landscapes. They spent much of the ride discussing what they would write. Henry drove, Kamila read the map, and when they got lost in conversation, they got lost on the road. Realizing their mistake hours later, they retraced their route, arguing as couples do.

Eventually, they reached the Kalahari and soon arrived at Henry's ancestral palace. Kamila couldn't help but laugh: it was just a manor house—albeit a big one, as Henry insisted. Next, they visited the huge mountain that young Henry had rolled down in a wheel. He didn't need any toys back then, he'd become fond of saying; he just squeezed his way into that wheel and rolled down the mountain until he felt sick. Well, that mountain turned out to be more of a hill—albeit, a rather high one, as Henry pointed out. Kamila snickered again despite her best efforts. A child's view of the world is so different from an adult's: everything is bigger, more colorful, louder, more adventurous.

Kamila quickly fell in love with Henry's native habitat. It was almost disturbingly breathtaking. It was the opposite of the urbanized Western world. The silence! It was a deep, tense, screaming silence. It was confusing because you also heard the sounds of animals—the buzzing insects, the rattling reptiles, the singing birds, the digging aardvarks—but their sounds blended with nature, were swallowed by it, were one with it, and got lost in the vastness among the grass and sand, the bushes, and the faraway Korana mountains.

This landscape doesn't overwhelm the eye. It has structure—a few thorn bushes here, a watering hole or a dune over there. It seems to be designed in accordance with a grand, even-keeled plan. The mud-sand huts with their rounded, thatched roofs fit in seamlessly. And the endless sky ties it all together. Kamila had never seen so many stars. The Milky Way seemed so close you wanted to run your fingers through it. This is how the world looks without light pollution, without cars, lamps, reflections—the Earth shrouded in perfect black except for the twinkle of the stars. Kamila began to understand why Henry didn't feel at home anywhere else. She understood how a microscopic glimpse of a neuron galaxy, which so resembled the sky here, had captivated him twenty years before. She understood why Henry had called his first daughter Linoy, beauty of nature, and why he had suggested coming here to write up their discovery. Only the written word can conclude scientific research; it is as important as the formula. Only the written word can get you to the heart of the matter, to the point, to the big picture.

"How lucky you were to grow up here," Kamila said to Henry. And how unlucky Kai had been not to grow up there.

* * *

They ventured into the Karoo, the most beautiful part of the Kalahari, driving over ocher-colored sand, looking out into the distance, enjoying the silence together. After two days, they arrived at a camel thorn tree, or a *kameeldoringboom*, as it's known in Afrikaans. Covered in long thorns, growing dozens of feet tall so that only giraffes can access its offerings, leaning over to snack on its leaves. Antelopes fleeing lion attacks like using small camel thorn trees as their fortresses. The thorns protect their sides; only the antelopes' horns stick out. Henry loved antelopes. As a child, he pitched stones in their direction—not to hurt them; he didn't throw hard—and the animals would hit them back, using their horns like cricket bats.

"Let's take a photo," Henry said. They unpacked the camera and took pictures of each other: her in the shadows, skin sunburned in a T-shirt and jeans, looking off into the distance. They put the camera away and studied the landscape together: the dunes covered in ocher, the yellow-green grass, the savanna on fire in the evening sun. They saw water holes in the distance. The Kalahari is rich in water, but you have to dig for it, then install a little windmill that shovels it up from the depths and spits it out into big puddles like the ones on the horizon, which draw animals—elephants, hippos.

After a while, they broke the silence. "What an unbelievably intense world," Kamila said. The colors, the impressions overwhelmed her. It was almost too much. That made her think of Kai. "This is how Kai must feel. He lives in an unbelievably intense world."

III
UNDERSTANDING

1

How Kai Saw the World

You hear it slam. You see it flash. You can taste the pain.

The baby is sleeping.
Mom opens the door.
Light falls onto the cradle.
She lifts the baby up, strokes his head.
Her skin is slightly rough from bathing him, so he doesn't catch a single germ.
She says the most loving words.
She disinfects her hands. And changes his diaper.
The baby cries.

You're asleep.
You hear it slam.
You see it flash.
Thunder echoes in your head. Light stabs your eyes.
The light fires into your tongue. You can taste the pain.
It's stamping toward you. Your whole world wobbles.
It yanks you upward, your scalp in scraping pain.
Her voice hurts, screeching you feel to your fingertips.
Your nose burns all the way into your head.

Your butt is in scraping pain.
You cry.

That was Kai's life as a baby.

* * *

As an infant, you are programmed by nature to absorb everything. There's nothing you can do about it. You absorb it all, even if it's bad for you. You stuff dirt and poisonous plants in your mouth when your parents are distracted; you run into the water or climb onto windowsills. You don't *want* to explore—you *must*. And unlike a toadstool or a burning candle, all the dangers an autistic child faces on its discovery missions cannot be identified by parents. They are not dangerous for other children, after all.

Their parents don't know better. Occasionally, they may realize that something isn't doing their child any good. The child cries when you comb its hair, cries when you dress it, and cries when you bathe it. The child covers its ears. The child is loud itself. It's trying to drown out the sounds: the talking, the pipes, the vacuum cleaner; the clanking of the heater, the clatter of the plates; the animal voices, the revving cars. And that's not to mention the white noise inside, the rushing blood, the throbbing pulse: the whole, unrelenting organ symphony that drives tinnitus patients mad. How could a parent understand that?

Imagine stepping out of a cave into the desert sun and you'll get an idea how an autistic child sees "normal" light. Imagine hearing all the above-mentioned sounds—all the time. An autistic child perceives the world with a sensitivity that you may only experience a few times in your life: say, if you arrive in Australia during midsummer after a sleepless thirty-six-hour flight.

Kai grew up with this dissonance, torn between his drive to explore and this permanent jetlag.

* * *

And then comes the day when the child starts to withdraw; when he turns away, wanders off, rejects everything. It occurs slowly, almost imperceptibly. He would be lucky to escape Earth altogether, but he's stuck here.

On the street. Surrounded by a few people, though to him that's a crowd. Surrounded by city noise, though to his ears it's the equivalent of an airport runway. The blaring sirens seem faraway to other people, but to Kai they're in the middle of his head. Engines rattle at the traffic light. Kai rears his head from all the stench and the noise. A delivery guy rushes by, and Kai jumps out of the way, people jostling him. He could escape into a café, away from the street, the noise, the masses, but god forbid the waitress in there is tapping her pen on her order pad, or a light in the hallway is flickering, or a nervous guest is tapping his foot on the floor. Your friendly neighborhood café is an autistic child's private hell. The people speak loudly, laugh loudly, slurp loudly; the ice crushes and cracks, and the coffee spits while it brews. Everything is spinning, including the threatening fan on the ceiling, until the child throws itself on the floor and covers its head, as Kai did in Switzerland to his parents' horror. He only had one wish: calm.

2
What Have We Done?

We say that autistic people lack empathy.
No, we lack empathy—for them.

Henry cried. The desert blurred before him. Thirteen years of searching. Seven years since the diagnosis. It had been a long, hard journey: Kai's eyes in the clinic, his silence in kindergarten, his tantrums, San Francisco, Lynda. After the diagnosis, with the mystery solved, he had continued to work in the wrong direction. When you ask the wrong questions, you get the wrong answers. Powerlessness, impatience. But who can be patient when their son is suffering? The search had become a hunt, and no one showed him the way. To the contrary, all the experts and textbooks had led him astray. How lucky he was that Kamila, another searcher, had joined him. And now it fell like scales from their eyes. They understood.

"If God created the world," Einstein once said, "his primary concern was certainly not making it easy for us to understand." A scientist can never be sure, can never be satisfied. Progress is slow: you sneak from hunch to insight. It's like one of those jigsaw puzzles that Kai was so much better at than his father. A scientist finds one right piece, then another, and can be happy if the two fit together, but you're always far from seeing the big picture. You keep searching, keep trying new pieces. It never stops. This time, however, though important pieces were still missing, they could see the big picture. This is usually an unforgettable moment for a scientist, but to Henry it was a mere sideshow. Something far greater had eclipsed it: after thirteen years of trying, he finally understood his son. It was as if he had just gotten to know Kai. If only his son had been there with him, he would have hugged him. The hunt was over. Peace filled him.

And just as they sat there, with the sun sinking before their eyes, one thought after another rose to Henry's head. Oh, that's why Kai had done this and that. But not just Kai. He began to understand other children, like his friend's daughter who suffers from autism. Every time she was supposed to shower, a drama unfolded. She resisted like a cat, scratching, biting, water-fighting. Her angry father scolded her: Can't you just take a shower? It's just a shower! Everyone showers! Don't make such a fuss! It's only water! But she wasn't making a fuss. The drops of water felt like hot needles to her, torturing her. And since, like many autistic people, she didn't talk, she responded with her hands and feet, trying to save her skin with desperate force. Was that so hard to understand?

Yes, Henry thought, that's what it's like for parents and their autistic children. They speak different languages. They experience the world in such a different way. And while these thoughts unfolded in him, and he was thinking of the other father, it came to him: And what did I do? What have I done to Kai?

Going to the movies.

The vacations. Every trip was one too many.

India.

Death Valley. Where Kai refused to walk another step. They had to carry him for miles.

Henry couldn't pursue that train of thought. He felt a deep sense of guilt. He felt the pain he had caused his son.

His awareness of Kai's struggle had evolved decisively. First, they understood it; now they could empathize.

Henry and Kamila realized that things were upside-down. Instead of dwelling on the supposed mind-blindness of autistic people, we should be discussing our blindness to their needs. Rather than talking about autistic people's flaws, we need to focus on society's flaws. "We say autistic people lack empathy. No, we lack empathy. For them."

It cannot be said often enough. "We say autistic people lack empathy. No, we lack empathy for them." Could Kai ever forgive them?

As they wondered about this, another terrible insight dawned on them. Perhaps Kai could forgive, but he certainly couldn't forget. Henry and Kamila knew all too well how the animals in the lab had behaved. Not only were their feelings amplified and their senses overdeveloped, but they also couldn't forget. Every trauma, every minor pain branded them.

Again, they thought of Kai, who at the age of five could snowboard better than Henry, and now, after several falls, refused to ever get on a board again; who foolhardily rolled down the slope in India and now hikes only with the utmost caution. Fear guides him through life and away from everything his father wanted him to experience. Every overwhelming experience is a bad experience. And you cannot make up for it.

"We only had one thought," Henry says in retrospect: "it's too late."

* * *

Though they stayed on in South Africa, their vacation was over. From then on, it was all about the article, typing and writing page after page in English and scientific jargon, supplementing their findings with numbers and tables, sources and studies. The intense world theory of autism. Translated into everyday language, the piece began like this:

Intense World Syndrome

> Until now, no theory could explain autism in its varied manifestations. We are proposing such a theory. It is based on cell research and behavioral science.
>
> Autism is multifaceted and ranges from disability to superability, with the latter being the exception. Most autistic people are considered stunted. Building on that assumption, established medical treatments have attempted to stimulate the brain.
>
> We believe the opposite is true. The autistic brain is not stunted—it is too powerful. It is excessively interconnected and stores too much information. Autistic people experience the world as hostile and painfully intense.

Henry and Kamila explain how they went about their research. They write about cells and molecules, about the cerebral cortex and the amygdala. They do not mention Kai.

The word they use most frequently in the piece, 221 times, is the prefix *hyper-*, meaning "above, beyond." And it's easy to see why. So much about the autistic brain is above and beyond the norm: how stimuli overwhelm them, how the cells interconnect, and how their brains can change.

Infants are programmed to seek out stimuli, to absorb the world. They imbibe breast milk and information, and both are vital to their future growth and development. Kai absorbed everything. And he did so through broader pathways than normal children. That can go well, for a while.

Until they start to feel overwhelmed. Only those who are well-acquainted with the child can see the change. There's something wrong with its eyes. Its linguistic development starts to lag. The child doesn't feel pain. The jumble in its head prevents it from sitting still.

No one saw Kai's suffering. He seemed happy. He was drawn to the lights in the hospital, sought contact with others as an infant, shrieked for joy in India as he rolled down the slope. All this was a festival for his senses. He matured in fast motion. The autistic brain matures—meaning grows—unusually fast. The more stimuli there is, the more growth. But a system that grows too fast is destined to fail. Every biologist, sociologist, and management consultant will tell you that.

The autistic brain eventually collapses in on itself. Exposed to all that overstimulation, parts of the brain get out of control. Some areas overheat; others shut off entirely. The brain becomes unbalanced. And since the world is what your brain makes of it, autistic people live in another world. Their world is limited and at the same time incredibly vast.

> Flooded with stimuli, the autistic person can only perceive the world in snippets. They focus on these snippets with excessive attention and a frighteningly good memory. This leads to savantism but also to social withdrawal and repetitive behavior.

This is important and points to a fallacy in many old studies. Scientists have often shown autistic people pictures of faces while observing the part of their brain that let them recognize those faces. If there was no discernible reaction, it was seen as supporting the scientist's conclusion that their brain wasn't working properly, i.e. autistic people suffered from a neurological deficit. It follows that one could cut out that part of the brain to see if they could live better without it. All that is wrong, says Henry. Many autistic people can't recognize faces because their brain is overloaded, limiting its application. If you show an autistic child the faces of its favorite superheroes, instead of human faces, the apparently dead part of their brain is suddenly animated. It sparkles and rejoices like our brains do when we run into a long-lost lover. The feeling is indescribably intense. The child's brain is so preoccupied with these sensations that it has no space for other faces, for people who aren't part of its world.

> If autistic people have a hard time dealing with other people, it is not because they cannot interpret their

> feelings or cues. Autistic people are neither oblivious to feelings, nor do they lack empathy. They just experience the world as so painful that they retreat.

Their wealth of emotion makes them seem devoid of emotion. Three areas of the brain are primarily affected: the frontal lobe, the neocortex, and the amygdala. In some cases, the autism is more in the frontal lobe; in others, it's more in the cerebral cortex. In some cases, the cells fire particularly wildly; in others, only a bit stronger than normal. All forms of autism can be explained this way—the first overarching explanation. The children who become the most reclusive have the most powerful brains. It's infinitely tragic: the kids who feel the most are the least capable of expressing it.

Kamila and Henry use the technical term *hyperplasticity* in their essay. Plasticity is Michael Merzenich's area of expertise, Henry's colleague from San Francisco. His theory that the brain can be changed by thoughts alone was long dismissed, until it was proven that the brains of sea nomads adapt to life on the water. Plasticity is the basis of all learning. Like the brains of these sea nomads, the brains of autistic people adapt. They block out dangers: the noisy world, us.

In 2017, ten years after Henry and Kamila wrote that a lack of eye contact doesn't imply a lack of interest, *Nature* magazine published a study by a neuroscience institute in Boston, a partner university of Harvard, that used new technology to inspect what occurs in the amygdala when autistic people look someone in the eye. The closer they look into the other's eyes, the stronger their brain reacts to it. The scientists write:

> Individuals with autism spectrum disorder often report that looking in the eyes of others is uncomfortable for them, that it is terribly stressful, or even that "it burns." Traditional accounts have suggested that ASD is characterized by a fundamental lack of interpersonal interest; however, the results of our study align with other recent studies showing oversensitivity. The results have potential clinical implications: during behavioral therapy, forcing individuals with autism to look in the eyes might be counterproductive and elicit more anxiety.

Even a look into happy, attentive eyes only arouses one thing: fear!

* * *

In their research, Henry and Kamila found themselves on the same path that Kamila had been treading since her time at university. No, not inductivism: they didn't just do experiments, tabulate results, and try to derive a law from them. Instead of trying to confirm their thesis, they tried to refute it. They tried to square it with reality. For two years, they tested their theory against other studies again and again. No mistakes now. Kai's future was at stake, not to mention their reputation. Their theory had to stand up to scrutiny and had to explain why other studies had reached different conclusions. They took the other studies seriously. It was inconceivable that so many great scientists, whose books they had read, whose lectures they had visited, could have been wrong about everything. The lab results of the most important studies had to be in line with their results. Kamila had read through hundreds of papers. Indeed, the data was consistent, but their interpretation was quite different.

Henry and Kamila's work was a revolution. It was also an affront: to the American Psychiatric Association, which classifies autism as a mental disorder in their famous DSM manual; to the adherents of the mind-blindness theory, which says that autistic people can't empathize; to the research scientists who surgically removed the emotional centers from monkey's brains; to the corporations, whose autism medication stimulates the brain. None of their data or numbers could undermine the Markrams' theory.

Henry and Kamila didn't question that autistic people have trouble later in life empathizing and communicating with people. They only objected to the popular assumption that autistic people lack interest in those exchanges, as well as the notion that autistic amygdalas underperformed and had to be stimulated. Autistic people are very interested in social contact, but unfortunately their amygdala works so hard that they avert their gaze to avoid blowing a gasket. Their brains do not need to be stimulated. They need to be calmed down.

The guardians of the old doctrine knew all too well that autistic people were sensitive—this had been stated many times. But they saw this as one among many symptoms, another hint that a child might be autistic. They didn't dwell on it further. Only years after Henry and Kamila's breakthrough did they start to place more emphasis on it. The American Psychiatric Association made sensitivity one of the core criteria for autism in their famous DSM manual.

The decisive flaw in the old doctrine was not its conclusion but its initial outlook, the emphasis on deficiency. If you're on the wrong track, you'll never make it to the finish line, no matter how far you run. What Kamila had learned in her time studying philosophy had been substantiated: nothing is objective. Expectations guide results.

Henry and Kamila had also started in the wrong place. They would have never arrived in the autistic world, if Kai hadn't shown them the way. Everything had been turned around. They started in the hope of bringing him into their world. In the end, he opened up his.

3

No Window Seat?

The airline hit Kai with a lifelong flight ban.
And somehow you could understand them.

"The office called. I have to return to Lausanne," Dad said. He would have to leave a day early. The holiday was over. Kai was visiting South Africa for the first time since Henry and Kamila's wedding. The country was so big and beautiful. He had never seen so many animals in his life. Even the light here seemed completely different than in Israel.

They were supposed to fly together. Now Henry had rebooked, and Kai would have to fly alone. That didn't scare Kai. He had flown from Tel Aviv to Geneva many times. At age fourteen, he was a frequent flier.

But now his flight was being rebooked, meaning he had to fly back earlier. This was not as he planned.

"This is a challenge," Dad had said with a serious expression. "You think you'll manage?"

"Of course," Kai had said. He wasn't a baby anymore.

On the ride to the airport, Kai already felt it welling up inside him. He was going to fly KLM—not El Al, as he was accustomed. He wasn't going to get the special menu he had preordered either. Oh well, he wouldn't eat at all then. That call from his dad's office had really gotten him in a pickle. Kai tried to keep his rage in check. Dad was always so proud when Kai flew alone. Kai didn't want to disappoint him.

At the airport, they pulled his luggage out of the trunk and his aunt walked him over to check-in. An airline rep tied a little sign to his suitcase handle, and it disappeared on the conveyor belt. *Don't lose my suitcase now!*

Kai received his ticket, his aunt waved, he waved back, and then he was alone. El Al would have dispatched an employee to escort him through the airport. *It would be fine.*

Kai stood stiffly in the security line: X-rays, being patted down. He doesn't like it when strangers put their hands all over him.

He proceeded through the halls. That had to be his boarding gate. He stared at the screen in disbelief. What did they mean, *delayed*?

Kai hates delays. And not in the way other people hate them; you know, where they tap their feet and roll their eyes. Kai despises them to the core. An unrest grew in him. So rude: he had arrived on time, after all. He'd known since yesterday that he had to be there by 3:00 p.m. The airline had known about the appointment even longer. They had had all the time in the world to be punctual. Many people relied on them.

We're only delayed by a few minutes, the woman at the gate assured him. A few minutes? What was that supposed to mean? Fifteen? Sixty? There's a difference, you know.

Kai waited, but his rage was building: at the airline rep by the gate, at his dad's office. Finally, the magic words sounded: "Please have your boarding passes ready." On to the plane. Now he just had to look forward to the flight. Sure, he was flying alone, with a foreign airline, hungry, on the wrong day, not at the time they had agreed on—but he didn't want to be like that. He would accept his lot.

The flight attendant smiled at him. Kai looked at his seat number. "That one," she said.

"Wait, what? No window seat?"

"I'm sorry," said the stewardess, "but it's a good seat, and maybe—"

"I always have a window seat," Kai said.

He always flew El Al, always got his preordered meal, always flew on the day he was supposed to and not a day earlier because his father's dumb office had called. The signals that had been tormenting his neocortex started to run hot in his amygdala. The rage that had been simmering for the past day quietly boiled over, bursting out of all his extremities. No way! You can't do this to me! Fucking phone call! Fucking father! Fucking airline! Fucking menu! Fucking seat! Fucking stewardess!

Kai lost himself. When he came to, he was standing back at the gate, and his suitcase was on its way to the baggage claim. He was hit with a lifelong flight ban. He called his aunt: "Um . . ."

Henry wasn't mad at Kai or the airline—he was mad at himself. He had been thoughtless, letting them call him back to Lausanne like that. It wasn't the first time he had underestimated how hard life with autism can be and how easily things can get out of hand. Poor Kai. Henry knew how rotten he felt. He always cried after he had one of his attacks and said how sorry he was. "That wasn't me," he would say. "I'm not sure who it was. I'm so sorry. I don't want to be a bad boy." It breaks your heart.

Henry: He feels guilty and miserable: about how he behaved, what outbursts he's capable of, the fact that he had a panic attack. He feels ashamed.

Kamila: He broods over those situations so much, for so long.

Henry: That helps him grow—self-reflection. We encourage him to think about it.

Kamila: Kai is a challenge. Arguments, disagreements—they're a matter of life and death to him. This causes you a great deal of stress as a parent: What did I do wrong? What could I have done differently? You can't stop thinking about it. And you worry, worry, worry—until you learn to just accept it as a part of life. As a parent, you always feel guilty anyway: What could I have done? I didn't react well there! I don't spend enough time with my kid! I didn't say the right thing! I did the wrong thing! That's how a lot of parents think. However, if you manage to put all that aside, children are just children, and that's how it is with Kai. He's just even more demanding.

Henry: Yes, he has really pushed us to our limits.

Kamila: On the other hand, there are all the great things you experience as a parent. All the things you wouldn't have done without your children. All the things you wouldn't have thought. All the things that enrich your life. Kai has enriched us unbelievably, changed everything: how we think about the world, how we think about people, how we think about the brain. And he also enriched the lives of his sisters.

Henry: He taught them to empathize.

Kamila: Kai is a gift to them. How does Kai see the world? Why does he behave the way he does? That's how the girls learned to see others, to empathize. We learned that as parents too, and our daughters have benefited from it. We try to empathize with them more. When they throw a tantrum, for example. Or if Olivia finds her clothes too itchy. If she changes her clothes three times in the morning, because things aren't the way she wants.

Henry: With her, everything has to be perfect. When she gets mad, she's almost like Kai. But she isn't autistic.

Kamila: Not at all.

Henry, *laughs*: I guess you learn to deal with normal panic attacks too. Our children are lucky. They have experienced parents.

Kamila, *laughs*: Lucky devils.

Henry: I was twenty-six years old when I first became a father. I knew nothing. Once you reach adulthood, you've been prepared for so many things, but no one teaches you how to raise a child.

Kamila: Kai taught us a lot.

Henry: What an unbelievable amount of sensitivity is required, for example. We are continuously stunned by how much he has taught us. We talk about it a lot.

Kamila: So much of that knowledge is theoretical, though. I can talk for hours about the Intense World Theory and the miracles that occur in autistic brains. But when your kid refuses to wear a particular pair of socks in the morning, you still think, *No, not again, not every morning.* It's easy to forget everything.

Henry: It's not about never making mistakes. Everyone makes mistakes. You will never do everything right. But you can learn to deal with them, and you shouldn't run away. You've got to think about what you did. Recognizing mistakes is how we develop. It's the key to parenting. Your child is a multifaceted being. You will never know your children entirely. We thought we could, but it's not possible. You can't teach them to always do the right thing. But you can show them how to think about themselves, how to deal with themselves and their mistakes. If you can reflect on yourself, you can change. If not, you don't have a chance. That's very hard for an autistic child. He almost looks at himself like a third person.

But pity wasn't all Henry felt when Kai was crying again after his latest outburst, saying, "That wasn't me." It also saddened him, because Kai's introspectiveness, the key to change, had developed at a snail's pace despite all assistance. In lieu of that, every one of their interventions was a Band-Aid solution.

Each family member had their own way of dealing with Kai's attacks. Kamila got loud sometimes. That helped. Kai came to his senses. Linoy achieved the same result by crying. Kali took Kai's hand and whispered his compulsions away. The ever-rational Henry reasoned with him.

Anat practiced the hardest and most virtuous emergency relief. "What helps, what really helps: go up to him and hug him tight. That's it. He sobs briefly and begins to calm down. I get so mad at him sometimes when he says something stupid. I don't want to hear it anymore, don't want to see him, and so I do the opposite: I walk away from him, go to the door, but that only makes him madder—that's when he really gets going. I know what the solution is, I know what he needs, but I'm not always able to give it to him. I get angry and don't want to be close to him, don't want to touch him. But we parents, the whole family, we always have to keep this in mind: always, always, always. That's what autistic people need: they need that embrace, that compassion. They need precisely this security, this protection, this warmth. I counsel parents with autistic children and always tell them: if you want to calm them down, you have to be there for them in that moment, and you have to love and hold them. However, they're very good at sensing whether you're doing it mechanically, artificially, or whether your heart is really in it."

4
Counterintuitive

To give your child a chance in the future,
you must slow down its present.

Remember Paul Ehrlich, who brought medicine into the twentieth century? He made a decision that would save a lot of lives: instead of healing diseases, he sought to prevent them.

Autism doesn't fall from the sky, as Henry and Kamila's research has shown. It isn't destiny—it sneaks in, it develops. That means one can prevent it. But how exactly?

In a perfect world, where Henry had endless knowledge, funds, and time at his disposal, he would proceed like this: First, he would seek to understand the genes linked to autism. There are more than two hundred of them. Second, he would try to identify the triggers, of which there are many: manganese, mercury, alcohol. Third, he would devise an emergency relief program, so that you can defend yourself even if you are genetically predisposed toward autism and it is triggered. There is a reason that autism often only develops after a few years.

Henry and Kamila didn't have time—they had Kai. There was nothing they could change about his genes and triggers at this point. And so, instead, they asked themselves: How can autism be reversed, ameliorated, prevented?

* * *

The first three years of a person's life are decisive for their development. Their very nature is shaped, their senses, their language, their intellect, their feelings, their behavior, their capacity to love. Whatever you miss during that period will be very hard to make up for later. In those first years, children experience so-called sensitive phases, which is what scientists call the time windows where a child can acquire those fundamentals easily or lose them forever. If, for example, a child doesn't hear any language during those years, it will never learn to speak.

During some of these time windows, a child's synapses double in number and learning becomes literal child's play. However, if the child isn't provided mental fodder, stimuli, those same synapses atrophy again. Parents, teachers, and neuroscientists would like to know when each window opens. Everything seems to have its best time: the learning of language or a musical instrument, the establishment of self-confidence and social skills.

Autistic people have to protect themselves. They can bear to assimilate the world only in snippets. For children, this is a tragic situation. When an autistic child closes its eyes and ears to protect itself, it is doubly tragic. For, along with the pain, it is also blocking out vital stimuli. It develops poorly. The behavior that protects it now destroys its future.

Plasticity is another important factor. Kai remembers exactly what shoes he was wearing and where he was standing when Henry told him that they had taken the wrong hiking trail and needed to check into another hotel. That trifle was traumatic to Kai: he's as unlikely to forget it as you are to forget 9/11. Kai experiences his personal 9/11 almost every day. And he can't cast any of them from his memory. Being unable to forget is torture. Forgetting is a form of liberation. It is the brain's sewage disposal. If you can't forget, you end up mired in the past. If you can't forget, you end up stifling your spirit.

Neglected or abused children often behave like they're autistic. They sit there frozen by fears, unable to look up at people or approach them. They, too, are trapped in old suffering, not given the opportunity to develop. That's why doctors used to blame parents for their child's autism, because superficially it so closely resembles the trauma derived from neglect and abuse.

* * *

Henry and Kamila conducted another three years of research, studying cells and behavior. In the end, they established a set of recommendations for autism prevention and treatment that contradicted the old teachings as much as their foundational insight, the intense world theory, had.

Autism develops in the unborn child. Every mother should be careful in pregnancy, should be wary of medications, environmental toxins, and alcohol. If she drinks just a small glass of wine during the period when the neural tube closes, when the brain is formed, she exposes a genetically predisposed child to a one-in-fifty chance of developing autism.

If autism is diagnosed, it must not be treated like a mental disorder. Such treatment makes everything worse. Stimulating the brain accelerates autism.

At the first sign of autism, an avoidance therapy should begin, and it should continue until the child has gone through all sensitive periods of brain development, until it goes to school at the age of six. "Such critical periods are often irreversible milestones," Henry and Kamila write. The outbreak of autism can be mitigated and perhaps even avoided if the child spends that time in the right environment.

Unlike a normal child, an autistic child should grow up in a world that's filtered, sheltered, and protected. Their lives should be calm and predictable. "No computer games, no television, no bright colors, no surprises. Surprises can be painful. They remain in memory, and memories shape your life. It is hard enough for normal people to erase their memories, and even harder for autistic people." They go on to say: "The brain needs to be calmed down, learning needs to be slowed, stimulation reduced."

It sounds hard. "Try telling a mother or a father, 'We have the following plan for your child: we will stagger its brain development and its learning,' " says Henry. "It's counterintuitive, but this will allow its brain to function later." To give your child a chance in the future, you must slow down its present.

If a child's impressions of the world are muffled and filtered until it starts school, Henry says, you will avert the greatest danger, which is that parts of the brain go into a lasting phase of overaction. After those first six years, the child should take control itself, deciding what is good for it. "It should occur on its own, with gentle guidance and encouragement. Support your child as you see fit—but in slow motion. Let the child set the pace. It needs that feeling of security," write Kamila and Henry.

In the context of behavioral therapy, the child should cautiously be exposed to stimuli, allowing it to withdraw if it feels any stress. The therapist should only introduce new forms of stimuli in short, controlled instances. The goal is to let the child get accustomed to the stimulus and to lower its sensitivity. This is the same recommendation a group of Harvard scientists would settle upon years later, in the conclusion of their study on autistic people's aversion to eye contact: careful habituation.

Do not give them medication that stimulates their brain. The brain of an autistic child needs to be calmed down, its performance reduced, its learning inhibited, its forgetfulness supported. Henry remains wary of all medication currently available. "The doctors who prescribe them have no idea what effect they have on nerve cells. It is an illusion to think that mental illness, regardless which one, could be treated easily and in isolation."

Treating older children and adults is more complicated, but their suffering can also be mitigated, their clinical signs reversed. The therapy is the same: calm the brain, foster forgetfulness, reduce fears and stress. Don't intervene in their rituals, whether they're counting the lines on the pavement, building a LEGO tower, or completing a three-thousand-piece jigsaw puzzle.

KAMILA: Rituals are the result of fear. The children calm down when they repeat the same actions. You harm them when you take away their rituals. According to our theory, you should partake in their rituals, get involved in them. Eventually, they will come to you.

HENRY: Autistic children retreat into a bubble, where everything is safe, there are no surprises, and all is under control. In severe cases, the bubble can be quite intractable. You have to get inside it somehow. You have to sit outside the bubble and wait. That's the difference with usual therapies. You don't tell the children what to do; you do not push them. You wait and support them. And this is challenging, but it's the only way. Just support them and let them take the lead. It takes time. You may feel that it'll never work, but when they take the lead, it does.

* * *

An excess rather than a deficit of feelings? Curb the brain instead of stimulating it? Participate in their rituals? This set off an intense series of debates. Simon Baron-Cohen, one of the scientists responsible for the mind-blindness theory of autism, graciously tweeted: "Fascinating personal perspective into Henry Markram's 'Intense World' theory of autism," and linked to a glowing article about Henry's thesis by autistic science writer Maia Szalavitz. His equally renowned colleague Uta Frith, the second mind behind the experiment with Sally the doll, did the opposite and coauthored a harsh critique of the Markrams' theory:

> Our particular concern regarding the Intense World Theory centers on drastic suggested treatments for individuals with autism, namely withdrawing stimulation

> during infancy. The Markrams do not merely hint at such interventions, but explicitly spell them out. Yet if the theory is incorrect, these treatments could be damaging. As studies of Romanian orphans have strikingly shown, insufficient stimulation and impoverished neuronal input in early development are damaging to children's social, cognitive, and emotional functioning.

Firth and her coauthor further criticized that the Markrams had published their conclusions too soon. While the Markrams' study acknowledged that their theory had yet to be substantiated by systematic studies with human test subjects, it already called for a "drastic change" in the treatment of autistic children. Frith concludes:

> We need to verify this theory before it can shape our perception and treatment of autism. Once this is done, we may well find ourselves with an intensely interesting proposal. For now, we remain intensely worried.

* * *

Henry: When we published the paper, we weren't exactly greeted with open arms.

Kamila: Someone said we wanted to lock children in a black box—a complete misinterpretation, the opposite of what we'd written. All we were saying is that they should grow up in a calm and structured environment. Every child needs stimulation, as much as it can bare without feeling overwhelmed.

Henry: That can't harm any child. That's why we felt comfortable making those recommendations.

Kamila: I think it's just a scientific debate. We wrote a paper that went up against mind-blindness and other theories that people built their careers on, that made them famous. They read the paper and attacked it. That's the way of the world.

Henry: In the meantime, our theory is being cited so often that people can't ignore it anymore.

Kamila: And more and more scientists are agreeing with us. Every time we read their studies, we say to ourselves, "Yay, another one!"

More than thirty studies published in the past few years have supported Henry and Kamila's thesis. A group of MIT scientists cited the Markrams in a recent paper about hypersensitivity and autistic people's longing for a predictable world. Unpredictable stress, they write, is "one of the key aspects of torture and leads to the development of anxiety, fear, and aversion." They criticize treatments for autism that are only geared toward relieving symptoms, which is most of them.

A team of research scientists and doctors from Boston and Cleveland worked with autistic children and found that, even in moments of calm, their brains process 42 percent more information than the brains of normal children. This, they concluded, is what prompted their social withdrawal. One of the professors wrote an e-mail to the Markrams after the release of the study. "Unfortunately, when we wrote the study, we didn't know how much it dovetails with your theory. Instead we interpreted our results in the classical framework: autism as a withdrawal into the self."

The old problem: The wrong assumptions inevitably lead to the wrong conclusions.

He went on to say: "But our press release honors your theory, as is merited."

Henry: I don't really care about the scientific debate. The important thing is that it helps. Autism isn't a flaw. Knowing that benefits autistic people, brings them a bit of peace. Many people write to me: autistic adults, and, of course, parents.

Kamila: There was the mother from Hamburg.

Henry: Yes. It's hard not to cry when you read her letter.

5
Letters

"Please help my wonderful son."

Dear Mr. Markram,

Please forgive my spelling mistakes. I live in Hamburg. My English isn't great. I am the mother of a wonderful four-year-old boy. His name is Elias. He's autistic.*

Elias's father wants to put him in a mental institution. He claims that I made Elias sick.

Please help my little son. He is wonderful. He laughs a lot. He has three older sisters who love him. Unfortunately, he can't spend much time in the same room with them. He starts crying. He can't be in any room for long with more than two people. Their voices irritate him. He hates going out, because of the noise. He can't play with other kids. When other children come to visit, he is riddled with fear. He wants to do the same things every day. He listens to the same music every evening—he loves Grieg. He doesn't understand what time is. He doesn't understand the difference

* Names have been changed in all the letters quotes here.

between breakfast and dinner. On rainy days, he cries, because he thinks the day is getting "old" and he has to go to bed soon.

He cannot be alone in a room.

Please, Mr. Markram, help my wonderful son.

Sarah

* * *

Allow me to introduce myself: I'm autistic and have an autistic son and a highly gifted daughter. I was able to arrange a special training for her. The trainer recommended a calm and predictable environment. Her room should be painted in a neutral color. It has only one shelf, and I only put two toys on it. Later, I added some building blocks, but very slowly.

I'm sure such an environment would have benefited my son, and me too. My daughter is five years old now, highly gifted, and though she exhibited sensory abnormalities in early childhood, they have since disappeared.

Thank you so much for your time,

Brandy

* * *

I have Asperger's and read your "Intense World Theory" with pleasure. Finally, people who don't have autism will find out what we autistic people have been trying to show (and say!) to them all this time! During my childhood (since I was very small), during my adolescence, and as an adult, I was told that I had few if any emotions, little to no sensitivity or any kind of perceptiveness. Essentially, I was told that I wasn't an "I" and that's how I was treated. I often told my parents, teachers, doctors, and therapists that I perceived things more strongly, not less strongly, than them. They chalked that up to bad behavior or to the lies of a feeble-minded person who refused to face her problems. I was punished for that "bad behavior." May I ask you, doctor: How do you think such an approach to education and therapy can affect a person's health? Post-traumatic stress disorder, I suspect. It's like treating anemia with blood-letting.

Kate

* * *

Dear Henry,

It's almost as if autism is a contagious and shameful disease. I work full-time, which apparently only 15 percent of autistic people do. I struggle to be understood in my workplace. The management sees my needs as an attack on them. After the diagnosis in 2013, I thought my life would become easier. It has not. Last week, I started cutting myself. I thought no one would ever understand me. I don't even know how you can help me, but I just had to write you. Who knows? Maybe I can help you, too.

Tears are streaming down my face as I write this. Attached find a photo of me, one where I'm smiling. I think the way I look misleads people. They can't see my condition, because I look like everyone else.

Sandra

And so it continues:

I am a speech therapist and I have a ten-year-old son with autism. I am very influenced by your work and treat my son and other autistic children accordingly. (Liz)

I read an article about you in the Süddeutsche Zeitung. It expressed exactly what I feel. I have feelings, not all the time, but so intense. (Sandor)

I have two nephews, autistic. I run marathons to raise money for them. They definitely have empathy. (Heather)

I am forty and was only diagnosed recently. Reading your theory, I felt that someone understood me for the first time. I experience the world as too fast, too loud, too intense. (Fabrizio)

I was very moved by the article about your son and agree with your thesis. Since I have autistic traits myself, I know the excessive demands of the normal world all too well. (Dagmar)

Henry and Kamila receive countless letters and e-mails from people with Asperger's and autism, parents and friends, academics and handymen, doctors and laymen. There is plenty of writing about the Markrams' theory online in blogs, forums, and comment sections. One such comment, by a mother, recalls Frith's indictment of their work:

> My world is completely changed. It was the first time that I'd heard anyone working in the autism field who did not speak of it as a deficit. I remember reading every paper they'd written, making my husband, Richard, read everything I was finding as well. We stayed up every night for weeks discussing what this might mean, how it changed our view of our daughter, Emma, and how it completely upended how we worked and communicated with her. I felt as though everything I thought I knew about Emma opened up and I was introduced to a vibrant, new, and hopeful world.

Uta Frith's concerns must be taken seriously. Henry and Kamila have indeed opened a door for autistic people and their families, those who have had to live with the word *incurable* for so long. Their theory moves affected people to their core, stirring hope and longing—a responsibility the Markrams sought as little as fame or money. They haven't earned a dollar with their theory, publishing it for free on the internet, and they sell neither medication nor treatment plans. They can't help the whole world, not the mother from Hamburg, nor Kate from London, nor the autistic father from Basel. They read e-mails, they listen, answer, and encourage. They didn't start with a grand plan. It was all about Kai. He somehow got them caught up in this. Now they can't change it—they wouldn't want to if they could—and must live up to the responsibility.

The studies must continue, for the sake of all autistic people and their families. Their findings require further confirmation and more tests on human subjects. Their correspondents expect that from them; some very openly, like Kate, who writes at the end of her e-mail:

"May I also ask you: what, if anything, can be done or at least suggested (by you or others) to remedy the psychological and neurological damage caused by a) the disorder itself and b) the wrong treatment?"

Their studies continued. Next, they would research how to help autistic rats recover. In the meantime, Henry pushed ahead with a project that would make all animal experiments unnecessary. It promised to be the biggest scientific mission of all: the simulation of the brain.

6
Building a Brain

> *When you visit the pioneers of the new world, like Jeff Bezos or Bill Gates, you meet characters quite similar to Henry.*

We live in an exciting and frightening time. It was chimed in by a technological revolution, like every new era. Humanity arguably underwent its greatest transformation when Johannes Gutenberg invented the printing press, giving everyone access to knowledge. People learned to read and then to write, and their way of thinking changed. This, in turn, changed how we did scientific research. In the Middle Ages, science was captured in pictures. Pictures can describe a lot at once. They depicted totalities. Writing cannot do that. Every aspect of a scientific concept must be explained, one after the other. Science became linear: it looked at specifics before arriving at a big picture.

One hundred and fifty years ago, the world experienced another great transformation. The steam and gasoline engine thrust us into the age of industrialization and economic pioneers. Cornelius Vanderbilt's railroad, John D. Rockefeller's oil, and Henry Ford's conveyor belt changed the pace of the world. Machines replaced labor. The number and clout of blue-collar workers declined, and their children became skilled white-collar workers. Medicine changed along with the rest of the world. Great minds like Robert Koch and Paul Ehrlich defined this new epoch.

At the beginning of our era, there was the computer. In 1962, Marshall McLuhan's famous "Gutenberg Galaxy" theory anticipated our electronic age, which he said would end both the Gutenberg era and the industrial era. Once again, entrepreneurs led the way. Like their predecessors, they are accumulating unbelievable wealth and frightening power. The pioneers and builders of this new world are called Steve Jobs and Bill Gates, Larry Page and Jeff Bezos.

Jobs and Gates saw this new world coming before anyone else. In the world they envisioned, there would be a computer in every house and every pocket. Page and Bezos, the founders of Google and Amazon, have since surpassed the Microsoft and Apple founders in terms of raw power, recognizing before anyone else how the internet would change our way of living and communicating, how it would set the pace of the world. Like Ford's assembly lines, their algorithms are changing everything. Just as machines once replaced our hands, they are now replacing our minds, with white collars now facing the same fate as their blue-collar ancestors. A study by Oxford University predicts that by 2030, half of the current jobs in the United States will be redundant. Science will change and so will medicine.

The *Süddeutsche Zeitung*, Germany's most prestigious newspaper, sent me to meet some of these builders of the new world: Bezos, Gates, and Twitter founder Jack Dorsey, whom *Time* magazine has anointed as the rightful successor to Alexander Graham Bell, the inventor of the telephone. All three of these men have much in common with Henry. They understand technology like few others, are socially awkward and exceptionally smart, believe unconditionally in progress, and intend to change the world.

When you walk into the lobby of Amazon's headquarters, you are greeted by a sign that reads: THERE'S SO MUCH LEFT TO INVENT. SO MANY NEW THINGS WILL HAPPEN. WE ARE IN NO PLACE YET TO FATHOM THE INTERNET'S IMPACT. IT'S STILL DAY ONE. The quote is from CEO Bezos, the richest person in the world.

"Jeff," says Bill Gates, "is on par with Johannes Gutenberg."

"Amazon," Barack Obama said, "is the twenty-first century."

Bezos has not only changed how we shop and how goods move around the world, but his corporation has grown into a global superpower. Nothing in the world can stop him, Bezos says. "What's happening to bookstores isn't Amazon. It's the future."

"We must realize that Amazon's model, how they use data, is part of a major social transformation," sociologist Colin Crouch says. Computers and data are the key to this new age. They combine the achievements of the Gutenberg age and industrialization: they are spreading human knowledge. They utilize the power of machines—their thinking power this time, artificial intelligence. Like in Ehrlich's time, these new tools will revolutionize medicine and help us cure diseases that once seemed unbeatable.

I recently traveled to Silicon Valley to meet Stanford professor Sebastian Thrun, the inventor of Google Glass and Google cars, an engineer, not a scientist. With the help of his Stanford colleagues, he recently invented an app that diagnoses skin cancer and apparently does so more reliably than seasoned dermatologists. A computer's eye and brain can compute so much more than we can fathom.

Henry saw the rise of technology in medicine coming early on. The kind of run-of-the-mill diagnosis he hated doing in medical school is one of the first jobs that computers will take away from doctors. Pioneers will find a way to make money off this development. To Henry, their real value lay elsewhere. He was convinced that mankind could not begin to cure the big disorders and diseases of the mind, be it autism, schizophrenia, or Alzheimer's until we began to understand the brain in its totality.

Henry explains: "We must understand the synergy between diseases. Autism has a lot of symptoms in common with other disorders. Thirty percent of autistic people have seizures like epileptics. What disease has most in common with autism? Is it Alzheimer's? Parkinson's? Depression? Psychosis? Migraine? No one knows, because no one knows how they're all related. Clinics all over the world have the data on this. You could connect them via a network. What works for one disorder also affects other disorders. Progress on any one of these diseases could advance the research on all of them. That's the accelerant we need. In Europe today, there are 30 million people with mental disorders. We have their data: their genes, blood values, brain scans, their complete case histories. If we could connect it, the state of our collective knowledge would change overnight. Of course, you have to consider data privacy. I have suggested such a project, but the scientific community doesn't want to hear about it. It isn't urgent to them, because they have no autistic children."

Scientists prefer to continue researching according to the established methods. This means that everyone's in their own lane, publishing their own data. When their research is complete, they write a paper that identifies them as the author, their careers advance, even if their work disappears into a remote archive. Every year, a hundred thousand papers are published in Henry's field alone. He couldn't read all those papers on synapses if he tried. How much better would it be if all that knowledge were shared, if all the data flowed into a single database? What if all that occurred before the experiments were even over, so dozens of scientists didn't end up doing the exact same research?

Naturally, it isn't enough to just collect the data and assemble it. That's the old, linear way of thinking. If you tried to build the brain that way, it would take hundreds of years. No, the computer would have to think in pictures, as scientists did pre-Gutenberg. Henry would have to simulate the brain, build it on spec, and improve it with the help of data as it becomes available. Like an incredibly difficult crossword puzzle, you wouldn't collect letters one by one, form one word after another, and plug them all in when they're complete. You start with the big picture, then you insert, consult, and correct. That's the only way to solve the mystery in our lifetime instead of leaving it for our grandchildren to figure out.

And yes, once the brain was simulated, in the not-so-distant future, he would help Kai.

* * *

It's July 22, 2009, a TED conference in Oxford. There are countless conferences in the world, many of them important, some of them meaningful, but only two are considered the gold standard: the World Economic Forum at Davos and the TED (Technology, Entertainment, Design) conference.

Observers who want to understand the status quo look to Davos. To the future-minded, TED is the measure of all things. Some of our greatest thought leaders have graced the famous stage: Jobs, Gates, Page, Stephen Hawking, Nobel Laureate in Medicine James Watson, neuroscientist extraordinaire Michael Merzenich. On this July day, it's Henry Markram's turn. The organizers have announced him as our "most promising frontiersman." The auditorium is full; Henry's first sentence already sends a murmur through the crowd.

> Our mission is to build a detailed, realistic computer model of the human brain. Why are we doing this? First, it's essential to understand the human brain, if we do want to get along in society. I think this is a key step in evolution. The second reason is: we can't keep doing animal experiments forever. We have to embody all our data, all our knowledge into a working model. It's like a Noah's Ark. Like an archive. The third reason is the two billion people on the planet that are affected by a mental disorder. The drugs they use today are largely empirical. I think we can come up with very concrete solutions how to treat them.

Henry takes a deep breath.

He was finally honoring the vow he had made after his sabbatical, to help Kai on a large and on a small scale. After years of preparation, in May 2005, he had launched his Blue Brain Project at the EFPL in Lausanne. Now he has fifteen minutes to explain it all. More than a million people will watch his remarks.

He begins, of course, with the brain.

> It took the universe eleven billion years to build the brain. The real big step was the neocortex. It's a new brain. Mammals needed it because they had to cope with parenthood, social interactions, complex cognitive functions. You can think of the neocortex as the ultimate solution. The neocortex is still evolving at an enormous speed. We ran out of space in the skull, and the brain started to fold in on itself, in columns. You can think of the neocortex as a massive grand piano with millions of keys. Each of these neocortical columns produces a note. When stimulated, it produces a symphony: your reality.

There is a new theory of autism called the "Intense World Theory," which suggests that neocortical columns are "super columns"; they are highly reactive and super plastic. Autistic people have a capacity for learning that is unthinkable to us. But if you have a disease within one of these columns, the perception is going to be corrupted.

Understanding these columns, Henry continues, would solve the mystery of what reality is—our reality and the reality of autistic people.

In 2007, the Blue Brain Project achieved an interim goal: simulating the neocortical column of a rat. A column is a miniature brain, which by itself can process blurred impressions of our environment.

For that simulation, Henry and his team had to re-create nerve cells. Mathematics helped them in that effort, specifically the equations of two Nobel Prize winners, and the supercomputer that could do hundreds of billions of calculations per minute certainly didn't hurt. One laptop performed the calculations of a single nerve cell. A network of ten thousand laptops made up the whole column, stacked to roughly the size of a refrigerator. A supercomputer for a single column. Henry uploaded the neurons, every single one, and watched them work. It was fascinating: without any help, the nerve cells became attuned to each other, sending signals back and forth. Their exchanges looked like swarms of birds, darting to and fro.

Henry had written his lecture only hours before his appearance on the TED stage. He only got nervous when he entered the auditorium and saw Cameron Diaz in the crowd. Henry looked down, sweat on his forehead. He had no way of knowing how exciting this project had yet to become. Four years later, the European Commission would anoint it as one of its flagship projects, pledging 1.2 billion euros of funding over ten years. They renamed it the Human Brain Project. In ten years, Henry declared at the official announcement of their victory, they would have simulated the human brain.

7

Fighting with Colleagues

"Yes, it's egoistic, you want to help your child, but you're also helping other children."

A French restaurant in Lausanne: a couple sit in a booth. He is slender, with thick brown hair, small eyes, and such a smooth, soft voice that she leans in to hear him. She has the roundest eyes, lending her face an almost childlike quality, but her gaze is bright and clear, and her voice is full and throaty. She flags down the waiter. Henry orders oysters with Tabasco, and a steak. She orders fish and a bottle of white wine. A long day is behind them. A babysitter is looking after Olivia and Charlotte. Kai's in Israel. Now there's time to talk, relax. Luckily, Kamila was on time today.

HENRY: That's the biggest source of disagreement between us. I have autistic traits myself. One of my compulsions is that I am always on time. Kamila, meanwhile, isn't exactly a stickler for punctuality. She's more Spanish in that respect.

KAMILA: Well, I did live in Mexico for two years.

HENRY: When you say 10 a.m., it usually ends up being 11. My brain is different that way. She has a Spanish brain.

KAMILA: Mexican, darling, not Spanish.

HENRY: When it comes to punctuality, you certainly don't act like that you grew up in Germany. If I have an appointment at 7 p.m., I'm there at 7 p.m. and not a minute later. Being a minute late is already unacceptable.

KAMILA: When we have plans, he's tense 24 hours before because he knows I'm going to be late. He's like Kai in that way. He is obsessed with the question: How do I get her to be on time? (*Laughs.*)

HENRY: Today, I had a doctor's appointment at eight in the morning. And I knew: there's not a chance she'll be on time. Why would she even make an appointment for eight a.m., then? So, at 6:30 a.m., I turn on the light, take a shower, make as much noise as possible so she wakes up—and still she's late. These are the things we fight about.

KAMILA: That's your autistic streak. When we met, Henry was very reclusive. He still is, in fact. It's in his genes.

HENRY: I'm more autistic than Kai in that way.

KAMILA: What do you do when you've been together for a while? You meet people. We kept to ourselves for a long time.

HENRY: We did socialize with people.

KAMILA: We visited my parents and my friends. And what did Henry do? Sat there and wouldn't talk at all.

HENRY: But then I would. I wouldn't let you talk. That was the phase, when you started kicking me under the table, so I would *stop* talking.

KAMILA: In everyday life, he is actually very sensitive. He's good at listening. But in the scientific context . . . At conferences, he's almost hyper. If you talk to him there, it feels a bit like you're talking to yourself.

HENRY: At work, I don't have a lot of sensitivity. In that context, I'm like a bulldozer. I set my goal, and everyone else better get out of my way. That's not sensitive at all. It's almost arrogant. "This is wrong, this is wrong, this is wrong—that's how I am then."

KAMILA: But not intentionally. In a naive way.

HENRY: No. I am aware of it.

KAMILA: You can't do that in business meetings.

HENRY: I had to learn that. I had important meetings for the Human Brain Project. I sat down and said, "Okay, here's our topic. Let's go." No "How are you?" No "How was your flight?" That wasn't part of my thinking. We weren't there to talk about each other, after all, but about the mission. However, particularly in Switzerland, you need to spend the first ten minutes building a rapport with people. Kamila had to coach me to sit down, calm down, ask questions, make small talk. (*Both laugh.*) That's another thing I have in common with Kai, but I've learned to do that now. He hasn't gotten past that point yet. He starts talking and just won't stop.

KAMILA: He does ask you how you're doing and what you've been up to, but after a minute he's done with that and tells you a story. And it doesn't matter if you fall asleep.

HENRY: Like him, I have a million tics. In school, I had reading and spelling difficulties. I mixed up letters. To this day, what I write is hard to decipher. But I developed coping strategies. It doesn't slow me down.

Unfortunately, Kai did not inherit that coping ability. Or he just didn't have his father's luck. Perhaps autism was never triggered in Henry because he received the right therapy, the Kalahari therapy, which perfectly lines up with what his research suggested: no television, no computer games, no bright colors, no noisy streets, no sirens, no movie theaters, no crowds—just the calm vastness of the savanna and its predictable cycles. The sun and the cows set the pace. It wasn't until Henry was ten years old, after his "critical periods," that a bit of movement came into his life, the rubber tires that propelled him down hills, the uncles who sat him behind the wheel, the bush pilots who took him to school in Durban. . . . Unwittingly, his parents and grandfather had done everything right. They had protected him from the noisy, fast world, had given Henry structure and endless possibilities to withdraw.

In the past, millions of people had the same good fortune. Though genetically they may have been predisposed to autism or another mental disorder, they just happened to grow up in the South African savanna or in the Texas boonies, in a world that naturally fulfilled all the recommendations of the intense world theory. Like Kai, millions of people are now unlucky enough to have been born into a noisy, fast world, which also finds its way into everyone's pockets. This may be enriching for ordinary people, but it is tragic for autistic children, who absorb everything until it all gives way.

Science is puzzling over why so many people in our time are autistic. One reason, surely, is that doctors have a better idea of what they're looking for. But the rise is too steep for this to account for it alone. It may be simply because so few children grow up with that natural backwater therapy.

* * *

When Lynda examined Kai, she also examined his family. Henry, she noted, "exhibits many characteristics of a hunter mind," which is what she calls Asperger's. One of the characteristics: "He is extremely attentive."

Henry was doubly lucky. After childhood, when he was ready and willing to get out, his parents and grandfather unknowingly continued this course of therapy. They challenged and encouraged him. His autism turned to his advantage. His super-columns catapulted him to the top of the scientific community. If Henry didn't carry this autism in him, he wouldn't have had the choice between elite universities like MIT or the EPFL. He wouldn't have developed the intense world theory, wouldn't be on a TED stage, and wouldn't have led the Human Brain Project.

When the TED producers announced Henry as a "frontiersman," they meant that he combines medicine and computer science. Henry is a frontiersman in another sense as well: he's at home in the world of autistic people as well as ours. This has its pitfalls. In her case history, Lynda wrote about Henry's ambition and what it meant for his family. When Henry helped his children study, it could never be fast enough. He overwhelmed Kai. Henry didn't wait for his son, and he lost him along the way. This, incidentally, is how he would lose many of his colleagues as the head of Human Brain Project. A top executive must also sometimes behave like a father and alleviate fears.

New ideas are always met with resistance, aversion—and the Human Brain Project was no exception. Eighty-five universities and hundreds of scientists were involved. Before long, two opposing camps formed within the massive project. Henry's people on one side; the traditionalists on the other. The latter wanted to do scientific research like conventional biologists: measuring brainwaves, tallying the results, and deriving laws. They considered Henry's plan fantastical. They resisted incorporating so much computer science into medicine. "Should Europe be spending one billion euros to support the passionate quest of one man?" one critic asked in *Nature* magazine. What if Henry was wrong?

The concerns were justified. Pioneers fail much more often than they succeed. The opponents, however, were ignoring an important point: the European Union had chosen this flagship project because they wanted to support a daring venture. A project in the traditionalist mold would never have received the grant money.

"Many were grateful to Markram for bringing in the money, but they nevertheless thought that they could sway the project in their favor," a scientist later told the *Neue Zürcher Zeitung*. These tacticians clearly didn't know Henry very well. They e-mailed him their concerns and received harsh responses. Their anger flared. "This is an IT project!" an opponent exclaimed.

Of course, it is. A scientist who thinks that's a bad thing hasn't understood the new era. If that person had lived in the late nineteenth century, he probably would have chastised Ehrlich for his "chemistry project!"

Amazon is an IT project, and it has changed the way we do business.

Facebook is an IT project, and it has changed how we communicate.

Airbnb is an IT project, and it has changed how we travel.

Google Cars is an IT project, and the car manufacturers fear it for good reason.

Classic neuroscience has existed for a century. It is important and will continue to exist for another century. But now the new wave demands its place. Biochemistry alone will not lead medicine into a new age—IT projects will. The question is not whether it will happen, but how. The traditionalists should consider this. Pioneers have a habit of moving like bulldozers. What benefits society in the long-term can hurt it in the short-term. Ford and Vanderbilt were not concerned about environmental protections or working conditions; that required unions and political action. And our new overlords are every bit as ruthless. Airbnb creates housing shortages. Amazon employees regularly report how overworked they are. Facebook shocks us with their data use. Google has paid billions in fines for unfair competition practices. Digitized medicine threatens our privacy. How does one protect the data of sick people—particularly the mentally ill and unstable, who have a long history of being stigmatized? How does one use clinical data anonymously? What other ethical questions does one have to ask, which the pioneers, the geeks, don't consider, because they only see progress, opportunities, rather than dangers, risks? These are questions demanding answers.

* * *

In an open letter, hundreds of scientists attacked Henry. His goals are illusory, they said, his leadership style authoritarian. They accused him of poor management. The criticisms were simultaneously right and wrong. Yes, his leadership style is authoritarian, his management style questionable. He should have held more conversations, been more considerate of people's doubts. He could have made more compromises, but that would have stripped his project of its radicalism, its risk, and thus of its meaning, its value. Indeed, the possibility of its failure was always great; but the promise of its success was immeasurable.

Henry didn't even consider the possibility of failure. In that way, he's wired like those Silicon Valley pioneers. They may have a lot of shortcomings, but they do understand a thing or two about making the seemingly impossible possible. Some years ago, I spoke to Ben Horowitz, who has a legendary reputation in Silicon Valley. The famous angel investor founded a company and later sold it for $1.6 billion. He writes a blog that reaches ten million readers. His business partner is one of the great internet pioneers, Marc Andreessen, founder of Netscape, who was considered the man of the hour at the outset of the internet revolution. The two manage billions of dollars, holding shares in Skype, Facebook, Pinterest, Airbnb, Twitter. In short, they have invested in everybody who is anybody in the Valley. From a small buy-in, they earned a hundred million dollars with Skype alone. Failure, Horowitz says, is the rule in the Valley. Ninety percent of companies crash; the other 10 percent fly, making him and his business partner rich. They seek out that risk.

Traditional investors are only concerned with past achievements and profits. Investors in the Valley, on the other hand, look to the future, the potential profit. If it's 1 percent likely that a startup will eventually make $100 billion, then—arguably—it's worth a billion dollars.

What is to be gained from the Human Brain Project? No money, certainly. It was about so much more.

"I demanded top speed. If someone came to me with a solution that took a long time, it was no good. Even if it was a great solution—that wasn't good enough. Of course, one can accuse me of doing all this to help Kai. Yes, that's egotistical. I wanted to help my child, but its success would benefit all children. What drives me crazy are people like that Nobel Prize winner, very well-known, very influential, who tells everybody that maybe, one day, his great-great-grandchildren will understand the brain. And he calls himself a neuroscientist. He might as well quit."

And so, Henry conducted himself like a Jobs or a Bezos, who were also known for harsh e-mails and furious zeal. But a Bezos could do things his way: his investors demanded it; his employees expected it. The higher the risk, the higher the possibility of everyone getting rich off company shares. By comparison, the world of committees, of professors and politicians, eschews all risk. Henry never had a chance. He was deposed as director. The project today does good, solid scientific research. Henry's dream is beyond its purview.

And yet: his entrepreneurial spirit, that first step he took, has made a difference. It encouraged China and the United States to do their own research in the same direction. Not content to let Europe leave them in the dust, they've started similar projects. Neuroscience is speeding on into the future.

Henry still leads the Blue Brain Project that took him to the TED conference. He still wants to simulate the brain, starting with the animal brain. But then! "I will make it."

8
Monica Is Crying

We want to find a cure, and that's wrong.

Tania moved on and became a professor in her own right. Monica Favre, a young graduate student, joined the team. Her professor back home at Duke University had taught Henry's work. Monica loved that seminar. She was a behavioral scientist and was fascinated how molecules and neurotransmitters can guide our behavior. During a visit to Switzerland, she paid the Markrams an impromptu visit. They talked, liked each other. "Why don't you do your PhD here?" Kamila suggested. Monica dropped everything and started.

"Henry and Kamila have incredible possibilities," she says. All the equipment, funds, and employees.

Yes, Monica thought, all would go well. She didn't suspect that she would soon be standing in front of Henry and Kamila crying her eyes out.

* * *

Monica was tasked with reintegrating autistic rats into general rodent society, mitigating and reversing their symptoms. The efficacy of her therapy would resonate in their cells and their behavior. Over the course of a year, Monica prepared the test series. Which rats would she choose? What medication would she administer? What experiment would she subject them to? Everything had to follow strict scientific rules.

She settled on this experiment: She would raise autistic rats in two cages and expose them to stimuli. But in one cage the stimuli would be predictable, and, in the other, surprising. What would that do to the animals?

Monica got to work. She first had to breed the rats. She administered an injection on the twelfth day of pregnancy, just as Tania had done. Only this time, the rats did not become autistic. Day after day, Monica sat in that incredible lab, equipped with "incredible possibilities," and didn't manage to produce anything even remotely credible. To be honest, she'd achieved nothing at all.

She gave the injections according to plan, as Tania had recorded it, but nothing happened; it seemed the rats had become immune to the drug. It went on like that for months—she worked thirteen hours a day—and she hadn't even managed to establish the prerequisite for her study.

Henry, Kamila, and Monica met on a weekly basis. At the beginning, Monica looked forward to these meetings. A weekly coaching session with a world-famous professor: What more could a medical student want? But over the course of those months, her mood shifted, and the meetings became torturous.

"So, what's new?" Henry asked every time, and she sat there, staring at the ground. Even her wild curls seemed to hang cheerlessly. "I was in a downward spiral. I didn't want to say anything or show anything to them. No data, no analysis, nothing. I went to the meetings and showed them nothing."

Henry let her work. He saw from the start that Monica was good. It wasn't her fault. "We have all been through what you're experiencing," he said. "Discoveries don't come easy. It's easy to devise a plan but realizing it can take some time." He gave her little tips here and there. "Try this, try that." Perhaps they were injecting the pregnant animals with another medication? The old brand no longer existed. Or they were procuring the rats from another breeder? The animals had another genetic structure. Such changes can influence the testing procedure. For the rats to become autistic, one had to inject the right dose at exactly the right moment. And the animal's exact genetic predisposition could factor in, as well.

Monica looked at Henry gratefully. "Yes, that makes sense. . . . Yes, I'll try that." She didn't want to give up that easily. Tania had also hit a rough patch in her research and ended up winning a prize.

Back to the lab, day after day, but the whole thing seemed jinxed. She tried everything, injecting a bit earlier, a bit later, injected a bit more, a bit less, waited for the animals to be born, but nothing, no sign of autism.

After a while, Henry and Kamila also started to worry. Kamila involved herself more, but her help didn't change anything. Before she knew it, another year had passed without the test series even starting. They had one last idea. A vet should examine the animals. And indeed, he ascertained that the rats had worms. So that was it. Parasites can affect animal behavior, their drives and fears.

"It was a catastrophe," says Henry. "We had to start over from the beginning." The animal house was cleared, disinfected over six months, and new animals purchased. The third year passed. All Monica wanted to do was cry. She had given up everything—for what? "I reached my limit. I started asking myself: Is this worth it? I worked thirteen hours a day, sacrificed my personal life. It got to the point where I said to myself: I don't need to be doing this. I promised myself: I'll keep going even if it goes nowhere. It would be all right not to finish my dissertation."

Once that was settled, having made peace with herself, she resumed the work, without pressure. She would try a few things she hadn't tried before, new techniques from molecular biology. She was still excited to try the experiment with the two cages. But meanwhile, her personal life took a turn; she became pregnant, took maternity leave, and then calmly divided up her time again.

* * *

Her doctoral examination drew closer—it was four months away—and she had nothing to show for it, but she refused to fret. She would do what she could. Kamila and Henry were supportive.

And then the day came when the numbers, the cells, and the rats in their respective cages started doing what Henry and Kamila had written five years ago. Monica could hardly believe it. She called her mother.

She asked her to visit for a few weeks and look after her daughter, freeing Monica to immerse herself in experiments, in data and molecules, and sit unshowered in the lab or kitchen for days on end, working in her pajamas, watching the puzzle take shape. All she needed now was time, time to transfigure this giant idea into a study that could withstand all scrutiny, providing evidence that the intense world theory had real world utility. Those were hard months. The weekly meetings continued, but now they were a pleasure. It was like Henry had said: discoveries don't come easy. He and Kamila were as excited as Monica. Henry's voice still lifts when he talks about those months in 2015.

KAMILA: Monica made a huge contribution.

HENRY: Her idea with the cages was great.

KAMILA: We raised the rats in two cages, big cages that they could play and interact in. They both grew up with more stimuli than the classic lab situation. Not much happens in a lab, usually.

HENRY: Rats are very social. If you stretch out a fine-mesh net in the cage between two specimens, they immediately meet in the middle, and sniff each other through the net. Autistic rats don't sniff each other; they avoid each other and prefer to play with wooden blocks. If you place them on an elevation, they get very scared and stop moving, while other rats will climb up anything and have no fear of heights.

KAMILA: Autistic rats don't like change. And things did change in our two cages: smells, food, and so on. In the first cage, these changes occurred in a constant and predictable way.

HENRY: And in the second cage, the changes came unexpectedly.

KAMILA: Like in real life.

HENRY: We introduced balls into their environment, which were easy to exchange. Another color, another size, another material. The little rats couldn't predict what would happen next.

KAMILA: We had the same setup in the second cage, but the changes were predictable because they occurred on a regular basis.

HENRY: We had to be very disciplined. We would introduce a new ball every Friday, a new smell every Saturday, a new wall in the cage on Sundays. And the results were spectacular. All their fears, all their feelings of confusion, were eliminated. They no longer avoided the other rats. They stopped playing with the wooden blocks and instead explored the other animals. And they did very well in our intelligence tests, for example, at recognizing and distinguishing between different sounds. They performed a lot better than before.

KAMILA: The same thing happened in the memory tests we did.

HENRY: Their strengths finally came to the fore. The more their weaknesses vanished, their fears, their rituals, the more their intelligence asserted itself.

KAMILA: The bottom line was that you can reverse the symptoms of autism.

HENRY: And prevent autism from developing. Monica's work has supported our theory.

* * *

And now? How will things continue in 2019 and after? Henry and Kamila are once again standing at a crossroads. Their theory is established, receiving more and more attention, and the research continues, but they are stretched to near capacity. They don't have enough time to conduct large-scale experiments with human beings and write more scientific papers. Their company, the science publisher *Frontiers*, is demanding; their other children, Olivia and Charlotte, are also around. Other scientists will have to build on their work, refine it, continue it. People like the research team at Harvard University, whose study demonstrated how important predictability is to autistic people. People like Ron Suskind, the Pulitzer Prize winner. His son, Owen, stopped talking when he was two years old and spent all his time watching Disney movies. It took Owen four years to break his silence. When his brother cried on his own birthday, Owen said, "Walter doesn't want to grow up, like Peter Pan," only to fall silent again.

Suskind understood that these movies could offer him a way into Owen's world. He sneaked into Owen's room, slipped a puppet of Iago, the parrot from *Aladdin*, over his hand and started a role-playing game. Owen found his voice again. Suskind wrote about it. In the course of his research, he visited the Markrams and asked them to explain what he had done instinctively right as a father. Today, he's the one giving talks and lectures to professors. He and his son seek to disrupt the clichés surrounding autism. He and the Markrams write each other every so often. They just got an e-mail from Suskind:

Holy Markrams,

There's joy in Suskindville. The Pope has designated us leaders of the global neuro-diversity movement—I'd submit, a central civil rights struggle of these times. After that meeting of Owen and Francis, our young champion was awarded a Pontifical Medal for his advocacy on behalf of the neurodiverse around the world.

Of course, it's our movement. The Suskinds and Markrams are both in the forefront. And we have more company and adherents than when we all conspired.

Suskind reports that scientists at Harvard and MIT have each started new tests, which seek to use artificial intelligence to get a deeper understanding of autism.

Harvard and the Suskinds: scientists working hand in hand with affected families. The Markrams are both. Naturally, they are still on the frontlines, as Suskind writes, while he carries Henry's message around the world. It sounds simple but can change everything: "Autistic people don't lack empathy. We lack empathy for them." This is a message that can be applied to many disorders and diseases of the brain.

"I have learned a lot from Henry," says Monica. "That there is another perspective. Real life. It's easy to lose sight of that in the lab. If you listen to what autistic people have to say, on YouTube, in blogs, when you hear that famous autistic woman, Temple Grandin, speak—they don't want autism to be cured. But that is precisely the first thing that we scientists want. We want to find a cure. And that's wrong. Meeting Kai and Henry was a turning point for me. The work I do in the lab is not about curing autism; it is about finding the biological reasons for its worst symptoms—the fear, for example. Find out where this fear comes from, if it's bad for the person, and then try to remedy it. Beyond that: let autistic people be as they are. Their brains are different, but we don't want the brain of every person to be the same. We just want them to give them a chance to be healthy, happy, and independent."

9
Pioneers and Child Prodigies

"Normies" didn't make our world what it is.
The others did.

How did we become who we are?

Since his college days, Henry has been fascinated by a particular answer to that question. It can be found in the work of Manfred Eigen, the great chemist, who won the Nobel Prize and then focused all his attention on the theory of evolution. His central insight seems complicated at first glance. It deals with genes and molecules, how life develops—living systems, as they're known scientifically—from the primordial soup to the one-cell animal, from ape to man. At its core, the answer is simple and possesses uncanny power. And it may inform our view of autism.

Living systems, including human ones, evolve slowly. They require game changers, pioneers, that take the whole species to a new place. There is a scientific word for these game changers that has an unpleasant ring to it: *mutations*. These mutations are the select few who change everything. They are different from the rest—abnormal. As such, they are initially considered disturbing to the system. They don't fit in. But in the long term, they are the ones who secure our survival.

These pioneers lead us into the unknown. According to the theory of evolution, most mutations die off. Some, however, survive and end up showing us a new, better way. They strengthen the whole species, developing it further. Without mutations, there would be no feathers and no webbing, no upright walking, and no intellectual achievements.

Societies are also a living system. They, too, develop through deviations. People were still dreaming of horse-drawn carriages while Henry Ford was busy inventing a car for the masses. When Konrad Zuse and a few other outsider nerds invented the computer, the head of IBM saw potential on the world market to sell around five of them. If everyone had always been the same, we would still be stewing in primordial soup. If our ability to mutate had faded, we would still be hunting, gathering, or farming. We would have stagnated. There would be no cars and no internet. "Normies" didn't make the world what it is today. The others did.

Kai is different. Autistic people are different. They represent an opportunity for the living system, for society.

A while back, Tania got a call from someone in the United States. The caller was an author who was interested in her and Henry's work and wanted to talk. The topic: child prodigies. The reason for her call was the work of professor Joanne Ruthsatz, a psychologist who had discovered that child prodigies and autistic people share a genetic mutation on chromosome 1. Whoever had it was predisposed toward autism or becoming a child prodigy.

Another study from 2012 found that half of all child prodigies are closely related—cousins, grandparents, siblings—to someone with autism.

Autistic people and child prodigies share so very much. They have an unbelievable memory, an uncanny eye for details, and they tend to develop insatiable passions. "It's so funny," Ruthsatz said in an interview with the *Huffington Post*. "Because when we talk about these traits in autism, we talk about them as deficits. But for child prodigies, they're strengths."

She suggests that scientists should study child prodigies to learn about autism. Whoever figures out why there are Mozarts, Einsteins, or Da Vincis will at the same time discover something about autism.

One might wonder how this aberration, which is spreading so rapidly, will change our society in the long run. That question is above our pay grade. The primordial soup couldn't imagine the single-celled organism. The ape couldn't predict the Neanderthal. The Neanderthal couldn't foresee the *Homo sapiens*. A human being living in Gutenberg's age couldn't fathom a person growing six feet tall and living to be a hundred years old. We have always believed that our generation was evolution's last trick, that our bodies and minds were fully developed. But our brains keep bridging new connections, growing new folds. In a way, this same process is now occurring outside of our bodies, in computers that we are teaching to think.

Perhaps autistic people are the vanguard of a world that thinks and feels differently. Perhaps they will remain the exception: one in fifty-nine, humans like us, just a bit different.

* * *

"Think of all they could contribute to society," Henry says. "If you raise them in a filtered world, you could end up with a genius on your hands."

This way of thinking has drawn criticism. Uta Frith bemoaned precisely this quote in her takedown of Henry's work. He puts parents under pressure, she argues, awakes unrealistic expectations, raises fears that they may be impeding their child's future. She quotes the president of a self-help organization, who says that one shouldn't judge an autistic person's worth by whether they have special talents.

No one needs to tell Henry that. He has a boy right in front of him who is not highly gifted. But to Anat, Kamila, and Henry, he is what every child is to their parents: the most valuable person in the world.

10
Great Expectations

I knew at that point that this was going to be the greatest disaster of all time. I knew it.

Kai has slept badly. His neck hurts. He yawns every ten minutes. He has a cold to top it off. Kai is in a bad mood. Lake Geneva glitters below, the Alps tower on the horizon. Springtime in Lausanne.

Kamila and Henry pick Kai up from the airport. How will the experience be this time? You never know with Kai. It can be lovely, funny, or loud and tearful. But one thing is certain: it never gets boring.

KAMILA: Remember Christmas?

HENRY: Oh, dear.

KAMILA: Kai loves Christmas.

HENRY: He started the tradition of us all singing together, the songs we sang when he was seven, eight.

KAMILA: Jingle Bells.

HENRY: Father Christmas. Whatever.

KAMILA: He starts singing each song. No one else is allowed to start. Otherwise, Christmas is over.

Henry: Yes, Christmas is over then. (*Laughs.*)

Kamila: The girls just watch, mesmerized. It's so funny.

Henry: And Kai plays Santa every year. One time, he was sitting in his room wearing his costume, and you could see that something was seriously wrong. "Kai, what's going on?" He didn't know how to say it: he had forgotten his Santa pants in Israel. (*Laughs.*) A very serious situation. We went shopping, got a whole new costume. Saved the day, last minute.

Kamila: He also hands out all the presents.

Henry: He usually panics before unwrapping his own gift. Last Christmas, we decided to give him a very special gift: an Oculus Rift, a virtual reality headset. When you put it on, it's like in a movie. You can be in a hang glider. When you look down, you see the ground below you, the woods, the meadows. It's exciting.

Kamila: Henry thinks he always needs to buy Kai the most amazing gadgets.

Henry: I always fall into the same trap. I buy the best gift, and it turns out to be a fiasco. Last year, I said, "Let's buy him a remote-controlled plane."

Kamila: That was not last year. That was many years ago. The first of many disasters.

Henry: A beautiful helicopter, huge.

Kamila: Like a drone.

Henry: No, drones are easy to fly. This helicopter wasn't. We went out into a field—

Kamila:—and when it finally got going, after all this back and forth with the batteries and the technology—

Henry:—the helicopter flies right into the forest.

Kamila: It just disappeared, after exactly ten seconds.

Henry: We spent a whole day preparing it and then never found it again.

Kamila: Kai was cursing. All we heard that day was "It's all your fault. You guys are stupid."

Henry: "You always buy me these dumb presents."

Kamila: Since then, I've been in charge of gifts.

Henry: The headset was my idea, though. The cream of the crop. It was a big package.

Kamila: Kai always gets the biggest present.

Henry: Of course, he said, "What is it? What am I getting?" And I made the same mistake I always do. I said, "You'll never guess." He responded: "Oh, yes I will." And then he spent an eternity guessing what it could be. Of course, he couldn't guess that it was an Oculus Rift. That wasn't in his repertoire.

Kamila: He mentioned everything: Wii, PlayStation—

Henry:—Xbox, everything. He had the biggest present under the Christmas tree, and he couldn't guess what it was. He started hyperventilating.

Kamila: I knew at that point that this was going to be the greatest disaster of all time. I knew it.

Henry: It had long ceased to matter what was in the package. His expectations had outgrown it.

Kamila: Perhaps we should give him an iPhone every year? We gave him that once, and it exceeded his expectations. Give him an iPhone every year, and everything's fine.

Henry: He can use that. When he's mad, he needs something to throw against the wall.

KAMILA: He's destroyed several iPhones that way. We didn't want to buy him one that year, because the last one had hit a wall. So, we said, "He'll get one next year again."

HENRY: So, we open the Oculus Rift. We had to get it to work.

KAMILA: A twenty-four-hour operation.

HENRY: On Christmas Eve, with two little girls. There's enough going on as is. You don't really have time to start an Oculus Rift. You first have to install the program on the computer. We tried that and realized that you couldn't install it on a Mac. So, I drove into the office to get a PC, the newest graphic card. We had everything one could possibly need.

KAMILA: Except a little thing called a driver.

HENRY: We realized this four hours later.

KAMILA: And Kai is standing behind you the whole time: "Is it working? Is it working?" And you always say, "Yes, Kai. Yes, soon."

HENRY: And I had to tell him to his face that I can't do it, especially not on Christmas. I had to call in a developer to solve the problem. Luckily no cell phones hit any walls. It was a miracle.

KAMILA: The magic word is *expectation management*.

HENRY: It does not change anything about the autism, but it's still important for everyone. A strategy for everyday life. You've got to think ahead and keep up with his expectations. When you don't fulfill them, when autistic people expect something and you suddenly do something else, it's a massive trauma. If you say we're going bowling at

7:00 p.m., Kai will start waiting at 4:00 p.m. In those three hours, he builds up such expectations, that if you think you can go bowling at 7:05 p.m., you have a problem. That will cause an explosion. It's all about expectation management.

Kamila: Sounds easy, but it's the hardest thing because it's the opposite of what we normally do.

Henry: Average brains can adapt. Can't find your shoes? Well, we'll be five minutes late then. That's a good enough reason. But not for them. To Kai, you had three hours, from 4:00 p.m. to 7:00 p.m., to find your shoes. You can save yourself the apologies.

Kamila: And because you think differently, you keep making the same mistakes. And sometimes you're so dumb that you even kindle his expectations.

They went to Venice last summer. Kai was already giddy with anticipation days before. Italy was the land of his favorite foods. All he could think about was pizza and spaghetti Bolognese. In a thoughtless moment, Henry and Kamila said, "Kai, we are going to buy you the best pizza and spaghetti Bolognese in all of Venice. It's going to be delicious. It's going to be a hoot."

And what a hoot it turned out to be.

They were traveling with friends, to make matters more complicated. After a long drive, they arrived at the hotel, one of the finest in Venice, five stars. "Where's the restaurant?" they asked the receptionist. Kai hadn't eaten all day. He had been waiting for his spaghetti since the day before and in the meantime had been stockpiling twenty-four hours of weapons-grade expectations. The fuse was lit.

Minutes later, they were sitting at a table with a thick white tablecloth and silver cutlery, surrounded by distinguished faces, busy managers, artistically inclined Venice Biennale visitors. The fountains spouted, the music murmured, the conversations did both.

"One spaghetti Bolognese," Henry said, before even looking at the menu. He was distracted by something the waiter couldn't possibly see: the burning fuse.

"Sir," the waiter said in a tone as fancy as his white shirt, "we are a vegetarian restaurant."

"Oh, my god," Henry let slip. Kamila's mouth gaped. And their friends, who knew what was going on, looked on as if they were about to be arrested.

"Vegetarian?" Henry whispered.

"Yes, the only five-star vegetarian restaurant in all of Venice," the waiter said proudly.

"Couldn't you . . ." Henry whispered, studying Kai out of the corner of his eye. Kai hadn't grasped the situation yet. In his mind, he was already spinning the noodles through the sauce, the most delicious Bolognese he'd ever tasted. And he had tasted a lot. It seems fair to assume that no person in the whole world had sampled more Bolognese than him.

Could they perhaps make an . . . ? His son had spent the last days . . .

"We have the most amazing grilled vegetables here. You will be surprised. Our guests fly in from all over the world because of our kitchen."

"Yes, of course. But, you know, children . . ." The waiter looked with some surprise at Kai, with his peach fuzz and his cell phone in hand.

"They can be a bit picky about food. Could you make an exception? The rest of us, of course, will gladly order your delicious vegetables."

"Well, unfortunately that's quite impossible. We don't have the ingredients. We don't even have spaghetti, and it's nighttime."

Henry took the waiter aside. Kamila started to prepare Kai; both their faces contorting when they heard what the other had to say. Henry warned the waiter: if he wanted to prevent a scene that he and his guests would not soon forget, he would be wise to give up on the menu and at the very least bring a pizza.

And indeed, at the last minute, a pizza came, a margherita, and Kamila and Henry topped it elaborately with words of consolation: it was the best pizza ever, and tomorrow, first thing, Kai would get his beloved Bolognese, the best in town, he would see, it would be better that way. And Kai sat there, looking down at his feet, his fingers in a restless twiddle, his teeth gritted. But then, miraculously, he somehow managed to maintain his composure, only mumbling, "Good, I won't eat anything then. I'll sit here and eat nothing."

"It was unbelievable," said Kamila. "After twenty-four hours of anticipation, even I would have lost my temper."

They were proud. They thought they'd made it to safety.

The next day, they hopped a vaporetto boat to the Biennale. The wind was blowing, the sun smiled, Venice smelled as it does, and a good-humored Kai looked out at the old buildings as they shrank into the distance and then grew again as they approached. He didn't care about the Biennale, needless to say. If it were up to him, a quick dash to the nearest restaurant would have been the first thing on the agenda, but Dad and Kamila had bought tickets long ago, so they decided to just eat there, in the sprawling food court next to the pavilions.

Then the scandal: no spaghetti Bolognese. Lots of sandwiches, curries, salads, cakes, all manner of things, but no Bolognese. Kai was beside himself. That was it. No Biennale, then. They jumped on the next boat, back to town, back to what mattered. They raced across cobblestones and bridges. Over there: that place behind the church looked good. Scanning the menu outside, they found relief: Bolognese, pizza. Kai ordered both. When the food came, he inhaled the vapors and ate the two dishes at the same time. Their three faces finally relaxed.

* * *

Here comes Kai. They spy him through the window at the gate. White T-shirt, blue chinos, white sneakers on his feet, four pearl bracelets on his wrist, a big silver watch, earphones in his ears, gel in his hair. He's wearing cologne. Seems like a normal, cool guy. Some people look at him a tad disapprovingly. A bit vain, they might think. They don't know better.

He wears these clothes because the fabrics are high quality, meaning they are softer on his skin. He loves the bracelets because Kali got them for him in the Philippines. He wears the medallion because his girlfriend wears the other half, and if you put the halves together, it says "Best Friends." He loves the watch because it's a gift from his father. And still, one has to admit there is a bit of vanity at play here. Poor Henry—who is gripped by horror when Kamila wants to take him shopping every six months—was dragged by Kai from store to store, examining two hundred watches, enduring two hundred sales pitches. Henry had already lost all hope when Kai finally said, "Yes, that's the one."

"He's the most spoiled child in the world," Anat sometimes complains to Henry when he has granted another of Kai's wishes, buying him a watch or cologne. Unfortunately, the cologne doesn't just smell of citrus. It also has a sad note. At first glance, Kai's no different from anyone else: he wants to look good and smell good so people like him. But it goes deeper than that. He wants to look good and smell good because he is—in the expert's jargon—hypersocial. He seeks human contact like no one else. And, of all people, he is the one whom people turn away from most. Because he is different. But maybe, just maybe, Kai thinks, if he smells good enough and looks good enough, people will like him.

Kai has a slender face, peach fuzz on his chin, black hair, chestnut eyes, and a kind of roundish nose you often see on teenagers as they go through their final growth spurt. He is twenty-four years old, but his nature, the way he feels and thinks, is ten years younger. He recently called his father from Israel.

"Dad?"

"Yes?"

"I would like to get an earring."

"What did your mom say?"

"She said I should ask you."

Henry was silent. This was an important moment. Better not screw it up. He had to think about this. He couldn't say no, but he also couldn't say yes. Eventually, he said, "Kai. You can get an earring. But I want you to really think it through before. Make sure that you don't just want it because others have one. Think about what's good about an earring and what's bad about one. Let's talk about it again this week. Then we'll decide."

I've done everything right, Henry thought, when of course he had done everything wrong. That week was a nightmare for Kai, not to mention Anat and Kali. The whole week, Kai did nothing but sit there and think and wait for the phone call.

"Kai, would you like something to eat?"

"No, I have to think."

"Kai, feel like going shopping with me?"

"No, I have to think."

And once he had decided:

"Mom, in six days I'm getting an earring."

"Mom, in five days I'm getting an earring."

He counted down so often that the whole family wished they could skip a few calendar days or at least fall into a deep multi-day stupor, just so they would never have to hear the word *earring* again. "Never set a deadline," Henry says in retrospect.

Kai got his earring. A while later, he called his father. "I have to tell you something," he said. "It's good that I had to think about the earring. Unfortunately, I didn't think about it enough."

It was a funny call. And it gave Henry hope. Autistic people are almost helplessly at the mercy of their compulsions. They find it hard to change their behavior, to steer it. It requires painstaking work. The prerequisite is that they even recognize their compulsions, that they recognize and nurture their weaknesses as well as their strengths.

These moments, when Kai reconsiders his behavior instead of seeing himself as a person without agency, are important.

They drive home from the airport. It's going to be a nice evening. They talk a bit more, but Kamila can't quite get through to Kai. He has discontinued his medication. She can feel it.

11
Needed

Kai isn't just tolerated. He's needed.
Society grants him his dignity,
and does itself the biggest favor.

The next day, Kai is sitting in the offices of Frontiers, Kamila and Henry's company. The sun warms the room, the pollen floats by, Lausanne shimmers. Smiling and chatting people drift in groups through the hilly streets. Kai's mood has lightened up, too. The employees greet him like an old friend, two dogs frolic around him. He is here often and likes it.

Frontiers' headquarters, located on a hill overlooking Lausanne, a stone's throw from the International Olympic Committee, are spread over two floors, featuring alcove seating, a kitchen, and a large open-plan office with a view of Lake Geneva. What started with Kamila's aggravation about not being able to download her own thesis has grown into an "open access" internet publishing platform. The idea is to share knowledge freely at no cost, for the benefit of experts and laymen. Parents of an autistic child, for example, should be able to divine the current state of research on the site. Kamila has won prizes for her work.

Frontiers publishes magazines, essays; 80,000 scientists help edit it, and 300,000 scientists contribute writing. Famous scientists publish their work on the site, while others use it as a basis for their research. No competitor—according to the well-known market researcher Clarivate Analytics—is cited in more essays about neuroscience and psychology. Frontiers is the number one source for authors and scientists. Frontiers has five hundred employees, offices in Seattle, in India, in London, and Madrid, and they are still hiring—scientists, programmers. Linoy works here too.

She isn't in this afternoon, and Kamila is sitting in front of her computer with red cheeks, way too much to do again, so Kai is hanging around alone, a phone in his hands, a plan for the day in his head. He's going bowling tonight. In Israel, he tells me, he goes bowling every evening at 7:00 p.m. What he doesn't mention: he is one of the best bowlers on his team and plays in the top league in the region.

It's easy to start a conversation with Kai. He smiles at you, looks into your eyes. He wants it to work. One can easily imagine him at his job. Kai works at the courthouse as a security guard. Experts have found that people with disabilities subconsciously affect others, making people they encounter more mild-mannered, more relaxed, more considerate. People with disabilities change the mood in a room without doing very much. This is a particularly valuable gift in a courthouse. They face upset, quarreling people—and calm them down. It's a job that couldn't be better suited to Kai. He isn't just tolerated. He's needed. Society grants him his dignity, and does itself the biggest favor.

This draws the situation elsewhere in the so-called first world—where people with disabilities are often still gawked at, stigmatized, and excluded—into sharp relief. It's inhumane, unreasonable, and we hurt ourselves as much as we do them. They shouldn't have to be highly gifted to get the place they deserve.

"Before I got the job in the courthouse, I sat at home bored," Kai says. "I woke up in the morning at 8:30 a.m., ate something, and then watched television. Every day was like that. Eventually, I watched TV until the next morning, until 6:00 or 7:00 a.m., because I had nothing to do. It was bad. When I started working at the courthouse, everything changed—that made everything good again for me. Now I have more fun. We have a lot of colleagues. And they love me. They really love me. They're happy to see me."

The worst day he's had so far was when he arrived twenty minutes late. Unforgivable. In the following weeks, he set his alarm clock earlier. He went to work at 5:00 a.m., leaving nothing to chance. But then he was so tired that he fell asleep at work.

How far is it to work?

Kai thinks about it. "It really depends," he says. "I can be there in three minutes or in five. I can take the long way, or, if I'm in a rush, a shortcut." Kai starts describing both routes, where he turns at the mall, what door that will take him to . . .

Kai doesn't do small talk. As easy as it is for him to start a conversation, he finds it that much harder to keep it going. He mumbles, loses syllables and words to the excitement, starts twiddling his fingers, plays with his keychain. He tries so hard, but he is on foreign terrain and accordingly feels insecure. He has met too many people that lose interest in him and turn away more or less politely. This is a shame, because he has such interesting things to say:

About a woman in his neighborhood, who can tell exactly how you feel when she puts her arm around you. She calms him down a lot.

About bowling, starting with his ball, which weighs fourteen pounds and is blue, yellow, and orange. And how you need to compose yourself, follow through with your arm. Most important, you have to clear your mind, not pay attention to anyone around you. "Don't listen to anyone. Just play for yourself."

About his grandmother, who died when he was a child. "She showed up at my kindergarten on my birthday." She never treated him like he was different.

About his worldview: "I feel things differently." He can't describe it exactly.

About his tantrums: "I was a bad boy. I hit and I spat. I was bad because I didn't know what to do. Now, I'm an adult."

About the boy in the neighborhood who pushed him: "I told him, 'I forgive you, but I will never forget.'"

About the school he transferred to, where the classes were small and the other children were nice. A boy there teased him. Kai didn't even look at me when he said, "My teacher saw it. She came over to me and said, 'I am really proud of you.' I liked going to that school. But I didn't like studying."

About his music: "It calms me down when I'm in a bad mood."

About his secret room in the basement, where he can hide and write songs—love songs.

About his dreams. He wants to be a professional musician, a singer. Or something with drawing. He's good at that.

About Fridays, when he cooks for his mom and Kali. Spaghetti, steaks, never fish or sweet potatoes.

About the best moments in life. When he sings Christmas songs with Olivia and Charlotte. "I start: 'You better watch out. You better not cry. Better not pout. I'm telling you why. Santa Clause is coming to town.' Then Charlotte sings, then Olivia, then it's my turn again. And everyone likes it."

About Kamila and why he liked to walk on the curb. Was he trying to annoy her? "No, I just wanted her to come and catch me."

About his sisters, whom he loves more than anything, and whom he envies, because they have something he doesn't. "They have a brother. Me. And I don't have a brother. I want one so badly."

About his parents' separation, when he was a little boy: "I was sad, but I always knew that my mother and father are the best of friends. I have friends who say, 'How can that be? Why do they still talk to each other? They're divorced!' I am proud of my parents. They're best friends. And my mother likes Kamila."

About his father: "I used to be a momma's boy. Today, I need my dad more. We go hiking. He's the perfect dad. When I pray, I thank God for giving me my dad."

A while later, Henry joins us. They talk about hiking. Once they went moonlight hiking. Fifty miles in three days through Switzerland. In the middle of the night, at 3:00 a.m., they arrived in a small town. All the hotels were closed. They hadn't expected that. "Let's sleep on a lawn," Henry said. Kai was scared. And it was raining. They went to a bar: Does anyone know of a room? Then a hotelier showed up. What luck!

"Are we going bowling today?" asks Henry. Kai nods. They talk about strikes and spares.

"Three hundred points is a perfect game. That's impossible," says Kai.

"What's your record?"

"A spare, five strikes, and then—"

"Then you messed up."

Kai laughs. "No, two more strikes at the end. Two hundred forty-five points."

"He's been beating me for a long time," says Henry. "When did you start beating me?"

Kai laughs again.

"I know how to beat you," Henry says. "I just make a lot of noise, distract you."

Kai laughs even louder. "Oh, yes."

"You're too sensitive."

* * *

It's 5:00 p.m.: two dozen employees meet up in the kitchen. The weekend has begun. They chat, laugh, pop corks. There are chips on the table, chocolate, celery sticks, and dips. Kai lounges by the counter, next to a small speaker. Any moment now, it will connect via Bluetooth to his cell phone. He's DJing today and plans to play Israeli pop and three of his own songs. No one pays too much attention when the music begins. The colleagues know Kai, and they let him do his thing, no excessive attention. That's good for him, he feels he's taken seriously. Being the DJ is a serious task. Without one, it's hard to enjoy your Friday night. An American guy opens his case, pulls out a saxophone. He wets the mouthpiece, listens to the music, and starts playing along with Kai's songs. One can see Kai straightening up, growing taller.

"When it comes to interacting with strangers, Kai is more spontaneous than me," says Kamila, as she tries to pry a dozen chocolate bars away from the already chocolate-smeared Charlotte. Kai listens to the saxophonist. A young colleague joins them, starts drumming on the bar, and the three of them band together, bobbing and dancing. Kai picks the songs and the others follow his tune. Eventually, Kai starts singing, in Hebrew, he sings other people's songs, then his own, about how he likes to go bowling with his father. The people keep talking, the music complements the mood, and when a song is over, the three performers high-five each other, the colleagues clap, perfunctorily, pausing momentarily. Henry gives a thumbs-up, Kamila smiles with her eyes, and Kai grows even taller. He walks over to his dad, who teases him because he only sings ballads now and hardly raps anymore. Kai leans on his father, laughs bashfully and plays with Henry's shirt buttons, too happy to stand still.

Time to bowl. They can't be late. Kai is so pleased that he lets Charlotte have his seat in the back. Playfully, he pinches her nose, tickles her stomach, and she puffs up her cheeks.

"The girls love Kai," says Kamila. "He brings color into their lives. When he visits, they hug him, kiss him, hang off his limbs. Of course, they have also experienced his tantrums, the strange behavior."

"It scares them a little bit," says Henry. "Though, if you asked them, I doubt they would say there was anything wrong with Kai."

"He's part of the family," says Kamila.

The bowling alley is in the middle of town. While the others are busy putting on their shoes, Kai is already standing in his lane with a ball. Kamila sneaks away with Charlotte, who can't stand the loud music, the screaming and cheering on the neighboring lanes. Kai throws like clockwork, strike after strike, but when it's Olivia's turn, he wheels over a helpful device: a ramp that looks like a dragon, which the little girl rolls the ball down in the direction her brother suggests. "Never look at the pins," he says, always at the dots on the ground and the arrows on the lane. Two rounds only, Kamila had begged, which leads to Kai taking Henry and Olivia's turns for them, playing against himself, though when he feels disturbed, he still sticks his white earphones in his ears. That's in keeping with his dictum: "Don't listen to anyone. Just play for yourself."

Once he's beaten himself, they take off for dinner. Turkish food. Kai has had his mind set on a hookah with apple tobacco since early in the morning, but now Lausanne's frenetic energy has turned against the Markram family. There are no seats outside, where it smells like apple tobacco, and candles dance in the midnight-blue hour. Hookahs aren't allowed inside, the air is thick enough, plates clatter, forks clutter, people talk at each other. Kai's face comes undone; this is not what he imagined.

"Let's go," Kamila says, with an uneasy face. "Let's go to the Italian restaurant, eat some pizza."

The girls hook arms with Kai, his face remains closed, but he does calm down a bit. The five of them stroll hand in hand through the pedestrian zone. Five hundred feet and they reach an Italian restaurant with empty tables outside. Everyone seems happy, except Kai. "I don't want pizza," he says, and orders ice cream instead, which he spoons silently while playing Pokémon Go. Occasionally he says something with a brooding undertone.

"Check, please," Henry says an hour later.

They go home hand in hand.

It's Kai's second victory over himself that day, more important than the one at bowling, and without earphones in his ears.

12
Beloved Kai

You shouldn't try to change anyone.

There was this girl. Kai knew her from school.

She was different. Not autistic—she was different in a different way. She sat there quietly, shy. When she looked at Kai, it warmed him.

When he left school, she remained in his head. Three years later, they met in town. She looked at him, and he walked over. "How are you?"

She showed him her bruises. She had a boyfriend.

Every time Kai ran into her, he asked how she was doing. And soon she didn't have a boyfriend, or bruises, anymore. Soon she had a new boyfriend, Kai.

She, too, found it easy to love Kai.

So, how does she look, Kai?

"She's a bit fat," he answers, "but I love her as she is. I say to her: You shouldn't try to change anyone. I love you as you are. In the beginning, she was a bit . . ." He makes a gesture: frightened. "Now she understands better. She has just got to be the way she is. And she loves me the way I am too."

Henry: It's a challenge, a delicate situation. One thing is clear: they aren't ready yet. In terms of his spirit and maturity, Kai is still a teenager. About fifteen, I would say.

Kamila: Thirteen, tops. He's like a child. Sure, he's matured in many ways. He's much calmer and has better control of himself than before. He doesn't throw as many tantrums. He tries very hard. When you say, "Kai, breathe, calm down," he withdraws and pulls himself together. But his emotional life and his mind are not like a twenty-three-year-old's. He's hitting puberty with all its problems. He is not only autistic, after all; he's also a person. Sometimes you forget that. Like everyone, he has good qualities and bad ones that he needs to work on. And sometimes you can be a bit inattentive with him. You think, *He's autistic*, and don't end up pushing him as much as you should.

Anat: As a mother, I make sure it doesn't get to be too much. It's difficult for him. He's experiencing situations that create all this noise in his head. He doesn't notice it himself. Or doesn't want to notice it. He just wants to show her his love. But I can see how it builds up in him. I'm there to give him permission: you're allowed to take a break. Or I tell her: Okay, don't call him for the next four hours. And you can see how he loosens up again. I hope that he realizes when it gets to be too much. I hope he makes sure not to get hurt and that she doesn't get hurt.

* * *

His first girlfriend!

Kai would take issue with that, of course. If you ask him, he's had plenty of girlfriends. Girls who talked to him. He even went on a date with one of them. She was blonde. Kai always wanted a blonde girlfriend. Anat liked her. "She would have been good for Kai—intellectually, too," she says. When Kai showed up with a new girl, she asked, as mothers often do, if he was sure about her. Didn't he want a blonde girl? "He looked at me and said, 'Mom, you have to look at the inside, not the outside.'"

He didn't feel good with the blonde girl. He wasn't sure why. Now, he knows, because he has something new to compare it to. It felt too cold. "Kai said to me, 'I've never felt anything like this.' He feels the warmth. She's a warm person."

And so, they sit there each Sunday, the girl sitting silently and hugging him the whole time, and he entertains and hugs her the whole time too.

"I sing songs for her, and I tell her stories," says Kai. "I love her. I tell her that every day."

Recently, Kai and Henry talked about it over a round of bowling, a man-to-man conversation. Kai told him that his girlfriend had warned him not to spend so much money. They would have to start saving. For a house. Kai looked very serious and determined.

"Yes, you always have to think about the future," Henry said. And he put his arm around him. His beloved Kai.

Acknowledgments

Thank you:

Kai, for letting me see the world through different eyes.

Henry and Kamila, for sharing your clever thoughts with me and letting me take part in your life.

Anat, Kali, and Linoy, for telling me so much about yourself and Kai.

You are a wonderful family.

Thank you:

Lynda Thomson, Monica Favre, and Tania Rinaldi, for your openness and scientific elucidations.

Thank you:

Skyhorse Publishing and my agent Kathrin Scheel. This book wouldn't exist without you.

Tony Lyons, for your trust.

Leon Dische Becker, it was rewarding discussing words and passages with you.

Cal Barksdale, for your sensitive editing and clever suggestions.

Thank you:

Susanna Benninger, for lending me courage.

Willi and Helga Benninger, for a desk in the most beautiful garden in the world.

Birgit and Horst Wagner, my beloved parents and first readers ("Write full sentences!").

Leonie, for the tranquility and reassurance.

And thank you, Franziska, for your love!